Numerical Methods:
Nonlinear Equations, Numerical Calculus, & Differential Equations

by D. James Benton

Copyright © 2018-2021 by D. James Benton, all rights reserved.

Preface

This is a combination of three books: *Nonlinear Equations*, *Numerical Calculus*, and *Differential Equations*. These three topics combine to form *Numerical Methods*. Nonlinear, integral, and differential equations are found throughout science and engineering across a wide variety of disciplines. These are a significant part of applied mathematics and much effort has been devoted to their study and solution. We will consider single and multi-variable problems. Various theories will be presented, but always with a focus on what works best—that is, *robust* algorithms.

The Finite Element Method

If you've struggled to understand the finite element method, then you must read this book. I don't cover a lot of theory in this text. It's mostly a compilation of examples. The one theoretical aspect of numerical methods for solving differential equations that I will present is the finite element method. This powerful technique is more often than not buried under a mountain of esoteric details that make it inaccessible to most students of applied mathematics. I aim to reveal, not obfuscate. I trust you will find this book helpful in that respect.

Programming

Most of the examples in this book are implemented in the C programming language. Some are contained in Excel® spreadsheets with macros. All of the program source codes (including a triangular grid generator and vectorized matrix solvers) are included. All of the programs will run on any version of Windows®.

All of the examples contained in this book,
(as well as a lot of free programs) are available at:
http://www.dudleybenton.altervista.org/software/index.html

All of the color figures can be found here (click on cover):
http://djamesbenton.altervista.org/

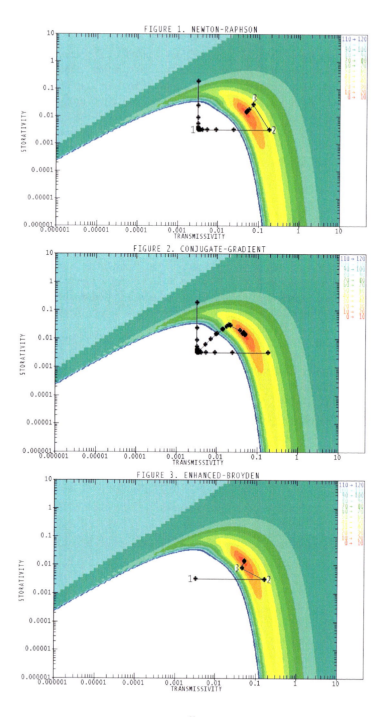

Table of Contents

	page
Preface	i
Chapter 1. Single-Valued Functions of One Variable	1
Chapter 2. Multi-Valued Functions of One Variable	14
Chapter 3. Functions of Several Variables	20
Chapter 4: Bounding & Scaling	28
Chapter 5: Nonlinear Least-Squares	30
Chapter 6: Quasi-Newton Methods	34
Chapter 7. Broyden's Derivative-Free Algorithm	42
Chapter 8. Approximating Gradients and Hessians	46
Chapter 9. Multi-Dimensional Bisection Search	50
Chapter 10: Evolutionary Method	56
Chapter 11: Nonlinear Regression	62
Chapter 12: Hybrid Regression	79
Chapter 13. Data Analysis	81
Chapter 14. Newton-Cotes Integration	85
Chapter 15. Gauss Quadrature	90
Chapter 16. Chebyshev & Lobatto Quadrature	96
Chapter 17. Composite Rules	99
Chapter 18. Green's Lemma	103
Chapter 19. 2D and 3D Integrals	106
Chapter 20. Improper Integrals	110
Chapter 21 Applications	116
Chapter 22. Differential Equations	151
Chapter 23. Explicit Runge-Kutta Methods	155
Chapter 24. Marching Method Applications	167
Chapter 25. Step Length Control	183
Chapter 26. Finite Difference Method	189
Chapter 27. Finite Element Method	208
Chapter 28. Boundary Element Method	225
Appendix A: Steam Properties - A Practical Example	237
Appendix B: Graphical Representation	246
Appendix C: Data Transformation Program	248
Appendix D. Newton-Cotes Coefficient Program	254
Appendix E. Gauss Quadrature Weights & Abscissas	257
Appendix F. Cooling Tower Demand	272
Appendix G: Generalized Runge-Kutta Function	281
Appendix H: Finite Difference Operators	282
Appendix I: Processing and Checking Finite Element Models	283
Appendix J: Solving Large Systems of Linear Equations	284
Appendix K: Triangular Element Generating Program	295
Appendix L: SIAM Paper	298

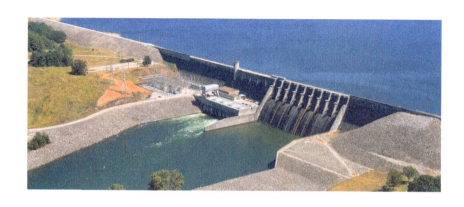

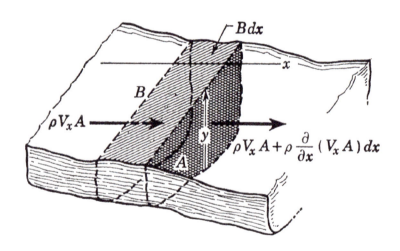

Chapter 1. Single-Valued Functions of One Variable

The simplest problem we can consider is that of single-valued, real functions of one variable—that is, $y(x)$ where y has one-and-only-one value for each x and there is only one value of x for any value of y. This excludes functions that wrap around, fold back on themselves, oscillate, or take on complex values, such as $\sin(x)$, $\cos(x)$, and $\sqrt{1-x^2}$. One such function is illustrated in the following figure:

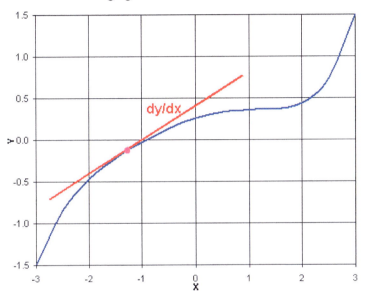

Most discussions presume you're interested in $y(x)=0$. For some other target value, simply subtract $y'(x)=y(x)-target$ and solve the modified problem. We will consider several methods for solving this problem.

Newton-Raphson Method

The Newton-Raphson Method begins with a guess (e.g., the purple dot in the preceding figure), the value of the function at that position, and the slope (or first derivative) of the function at that point.[1] The next guess (or *provisional value*) is calculated from the current one by taking the difference between the target and current value of the function and dividing this by the slope.

$$x_{n+1} = x_n + \frac{y(x_n)}{\left(\dfrac{dy}{dx}\right)_{x_n}} \qquad (1.1)$$

---

[1] Named after Isaac Newton and Joseph Raphson, was first published by John Wallis in "A Treatise of Algebra both Historical and Practical," in 1685.

This method works well some of the time, but not always. It may converge rapidly... or not at all. If the root (i.e., where $y=0$) lies in a region of x where dy/dx is small, changes sign, or vanishes (e.g., near $x=1.33$ in the preceding figure), you've got a real problem, because the iteration implied by Equation 1.1 will diverge (i.e., *blow up*). This is why you never want to use the raw, unqualified Newton-Raphson Method to solve any routine problem—you can't depend on it always arriving at a valid answer. The illustration we will use for this is solving the properties of steam (or a refrigerant) for unknown density, which you may need to do many, many times in a spreadsheet or other model. Here's what the iterations look like in Excel® (sorry about the ratty graphics, but it's the best you can get out of Microsoft®).

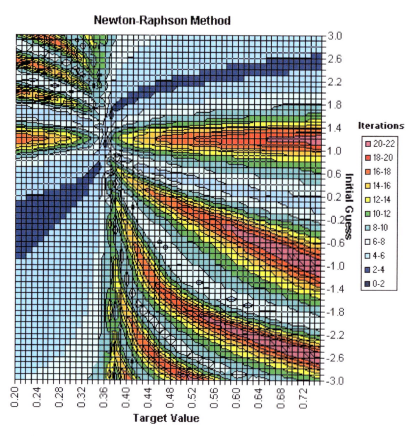

The Newton-Raphson Method always finds a solution with 61x56=3416 cases. The average number of iterations for this example is 9.4. The dark blue curving region (corresponding to zero iterations) is the shape of the curve, y(x), when the initial guess is equal (or very close to) the final solution. The red and magenta regions are where it takes 20-22 iterations to arrive at a solution to the specified tolerance (0.000001). If you're lucky and your first guess is close to the final solution (i.e., in the blue areas) this method is quite efficient. If you're not so lucky (i.e., in the yellow, orange, red, or magenta areas), this method isn't efficient. At least it always converges for this example—something that is not always true. The Newton-Raphson Method in VBA® is:

```
Function NewtonRaphson(target As Double, guess As
   Double, convergence As Double) As Integer
Dim x As Double
x = guess
NewtonRaphson = 1 'counting function calls
While True
  If (Abs(Y(x) - target) <= convergence) Then Exit
  Function
  NewtonRaphson = NewtonRaphson + 1
  x = x + (target - Y(x)) / dYdX(x)
Wend
End Function
```

Secant Method

We will next consider the Secant Method. The Newton-Raphson Method required knowledge of the derivative of the function. This is not always available. It may even be quite problematic to estimate the derivative. For example, where would you get the derivative of a stock price on the exchange? In such cases, you only have previous values of the function to work with and must come up with a *derivative-free* algorithm. One of the most basic principles of calculus is:

$$\frac{dy}{dx} \approx \frac{y_2 - y_1}{x_2 - x_1} \qquad (1.2)$$

and so we use this approximation to formulate an iterative step:

$$x_3 = \frac{x_1 y_2 - x_2 y_1}{y_2 - y_1} \qquad (1.3)$$

here the subscripts 1, 2, and 3 indicate the previous two values and the next step, respectively. It should be clear from Equation 1.3 that this method will run into trouble whenever $y_2 \approx y_1$, which happens more often than you might think. You must also come up with something to start this process off (i.e., the first step). The results for this problem are shown in this next figure:

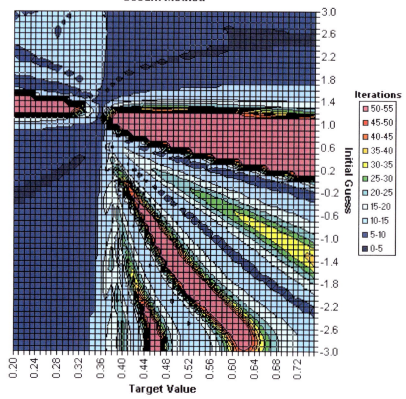

Don't presume that the Secant Method has a larger blue area (sweet spot for convergence) than the Newton-Raphson. In fact, the Secant Method performs so much worse for this problem that the scale has been increased from 22 to 55 iterations. Even worse that that is the fact that the values greater than 54 (i.e., the magenta regions) no solution is found, that is, the method diverges, rather than converges for 13% of the cases. When it did converge the average number of iterations required was 12.9, making the Secant Method a real loser. Its only value is historical. Don't use it for anything. The Secant Method in VBA® is:

```
Function Secant(target As Double, guess As Double,
    convergence As Double) As Integer
  Dim x1 As Double, x2 As Double, x3 As Double, f1 As
    Double, f2 As Double
  x1 = guess
  f2 = Y(x1) - target
  x2 = x1 - f2
  Secant = 2 'counting function calls
  While True
```

```
    f1 = f2
    f2 = Y(x2) - target
    If (Abs(f2) <= convergence Or Secant > 53) Then
   Exit Function
    Secant = Secant + 1
    x3 = (x1 * f2 - x2 * f1) / (f2 - f1)
    x1 = x2
    x2 = x3
  Wend
End Function
```

Regula Falsi Method

The next method we will consider is the Regula Falsi (Latin for *False Position*). Ideally, you want to always bracket the solution, that is, have one point on either side. With Regula Falsi, you may replace either the current or the previous step with the update, to achieve this. Regula Falsi is illustrated in the following figure:

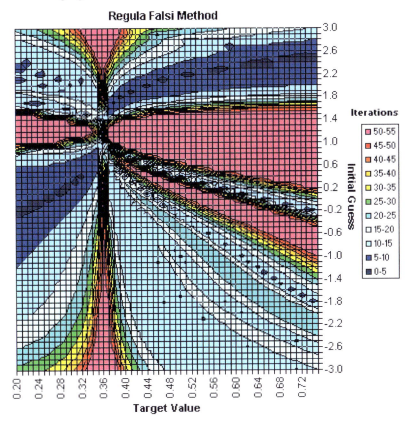

Notice that the regions of non-convergence (magenta areas) in the previous two graphs are different and substantial, making both of these methods a poor choice for applications. This method failed to converge for 19% of the cases and the average number of iterations required to converge when successful was 17.4, making this worse than the Secant Method. Regula Falsi in VBA® is:

```
Function RegulaFalsi(target As Double, guess As Double,
      convergence As Double) As Integer
   Dim x1 As Double, x2 As Double, x3 As Double, f1 As
      Double, f2 As Double, f3 As Double
   x1 = guess
   f1 = Y(x1) - target
   x2 = x1 - f1
   f2 = Y(x2) - target
   RegulaFalsi = 2 'counting function calls
   While True
      If (Abs(f2) <= convergence Or RegulaFalsi > 53)
      Then Exit Function
      RegulaFalsi = RegulaFalsi + 1
      x3 = (x1 * f2 - x2 * f1) / (f2 - f1)
      f3 = Y(x3) - target
      If (f2 * f3 < 0) Then
         x1 = x2
         x2 = x3
         f1 = f2
         f2 = f3
      Else
         x2 = x3
         f2 = f3
      End If
   Wend
End Function
```

Illinois Method

Both the Secant and Regula Falsi methods often overshoot. The Regula Falsi Method may also get bogged down on one side or the other, always updating steps on the same side of the root and significantly slowing—even preventing—convergence. These observations provide motivation for somehow adjusting the updates. One such modification is known as the Illinois Method, published by Ford.[2] There are two variants. The first variant is:

$$x_3 = \frac{\frac{1}{2}x_1 y_2 - x_2 y_1}{\frac{1}{2} y_2 - y_1} \tag{1.4}$$

Results for the first variant of this method and the same problem.

---

[2] Ford, J. A., "Improved Algorithms of Illinois-Type for the Numerical Solution of Nonlinear Equations," Technical Report CSM-257, University of Essex Press, 1995.

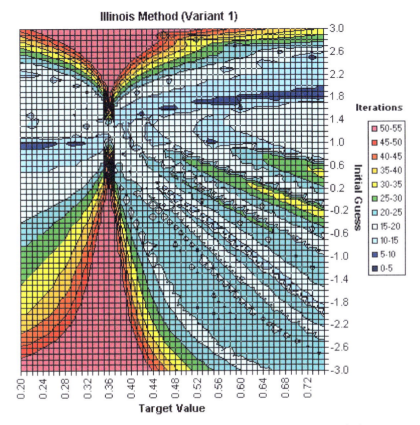

This method failed to converge for 11% of the cases and the average number of iterations required for convergence when successful was 22.8. The first variant in VBA® is:

```
Function Illinois1(target As Double, guess As Double,
   convergence As Double) As Integer
  Dim x1 As Double, x2 As Double, x3 As Double, f1 As
   Double, f2 As Double, f3 As Double
  x1 = guess
  f1 = Y(x1) - target
  x2 = x1 - f1
  f2 = Y(x2) - target
  Illinois1 = 2 'counting function calls
  While True
    If (Abs(f2) <= convergence Or Illinois1 > 53) Then
    Exit Function
    Illinois1 = Illinois1 + 1
    x3 = (x1 * f2 / 2 - x2 * f1) / (f2 / 2 - f1)
    f3 = Y(x3) - target
    If (f2 * f3 < 0) Then
```

```
         x1 = x2
         x2 = x3
         f1 = f2
         f2 = f3
      Else
         x2 = x3
         f2 = f3
      End If
   Wend     End Function
```

The second variant is slightly different:

$$x_3 = \frac{x_1 y_2 - \frac{1}{2} x_2 y_1}{y_2 - \frac{1}{2} y_1} \qquad (1.5)$$

Results for the second variant is shown in this next figure:

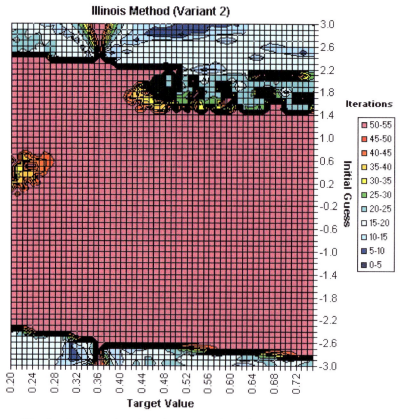

Clearly, the second variant doesn't work at all for this problem so there's no point listing the code that only differs on one line.

Anderson-Björck Method

Another modification has been proposed by Anderson and Björck.[3] This is similar to the Illinois variants, except the ½ is replaced by a factor m, as follows:

$$m = 1 - \frac{y_3}{y_2} \qquad (1.6)$$

If m is not positive, then it's set to ½ as before. The results for this method are illustrated in the following figure:

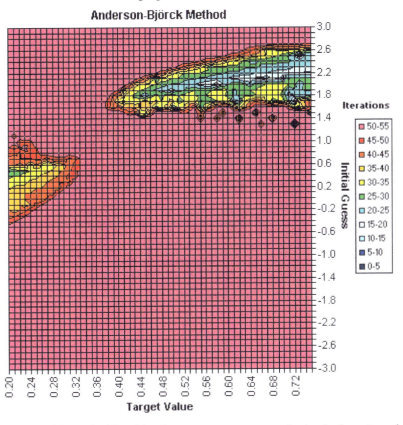

While this method is said to have won a contest, yet it clearly doesn't work for the current problem, failing to converge for 86% of the cases.

---

[3] Dahlquist, G. and Björck, Å., *Numerical Methods*, pp. 231–232, Dover, 1974.

Dekker's Method

Dekker and Hoffmann proposed that the Secant Method be modified to avoid dividing by zero in the case of $y_2 \approx y_1$, by replacing Equation 1.3 with $x_3 = (x_1 + x_2)/2$.[4] This is somewhat of an improvement, as illustrated in the following figure:

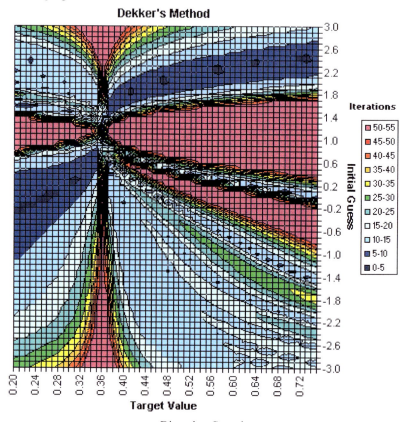

Bisection Search

While there may be some problems where these methods are adequate, for general application, they're useless because not finding a solution is simply not an option! The Newton-Raphson Method can be constrained and tweaked enough to work for some very special problems, and we will present one of those next, but first, the method that makes all of the previous superfluous: bisection search. It works all the time no matter what! It also converges every single time, regardless to $1/2^n$, where n is the number of bisections. It couldn't

---

[4] Dekker, T. J. and Hoffmann, W., "Algol 60 Procedures in Numerical Algebra," Tracts 22 and 23, *Mathematisch Centrum Amsterdam*, 1968.

be any more simple. Of course, it always takes 32 iterations, but that isn't a problem most of the time.

```
Function Bisection(target As Double) As Double
  Dim iter As Integer, x1 As Double, x2 As Double
  x1 = -3
  x2 = 3
  For iter = 1 To 32
    Bisection = (x1 + x2) / 2
    If (Y(Bisection) < target) Then
      x1 = Bisection
    Else
      x2 = Bisection
    End If
  Next iter
End Function
```

A practical application of the Newton-Raphson Method and comparison to the bisection search may be found in Appendix A. This book is not about making pretty pictures. It's about solving practical problems. I have provided a program to make pretty pictures in Appendix B.

Brent's Method

The desired solution is not always a root or zero point. We often seek a minimum or maximum. The preceding methods only work for functions that cross over the y-axis (i.e., have both positive and negative values). Brent's method searches for a minimum.[5] Where the Secant Method arises from a linear interpolant, Brent's method arises from a quadratic interpolant (i.e., the second order Lagrange interpolating polynomial). The iteration begins with three points (x_1, x_2, x_3), presumably bracketing the minimum. The function is evaluated at these three points (y_1, y_2, y_3). The resulting polynomial is:

$$y = \frac{(x-x_2)(x-x_3)y_1}{(x_1-x_2)(x_1-x_3)} + \frac{(x-x_1)(x-x_3)y_2}{(x_2-x_1)(x_2-x_3)} + \frac{(x-x_1)(x-x_2)y_3}{(x_3-x_1)(x_3-x_2)} \quad (1.7)$$

Take the derivative of Equation 1.7 with respect to x, set this to zero, and solve to obtain:

$$x = \frac{1}{2} \frac{(x_2^2-x_3^2)y_1 + (x_3^2-x_1^2)y_2 + (x_1^2-x_2^2)y_3}{y_1(x_2-x_3) + y_2(x_3-x_1) + y_3(x_1-x_2)} \quad (1.8)$$

---

[5] Brent, R. P., *Algorithms for Minimization without Derivatives*, Chapter 4: An Algorithm with Guaranteed Convergence for Finding a Zero of a Function, Prentice-Hall, Englewood Cliffs, NJ, 1973.

If the revised estimate is to the left x_2, swap out x_3; otherwise, swap out x_1. Continue until the improvement or the step size is less than some tolerance. A typical problem is illustrated in the following figure:

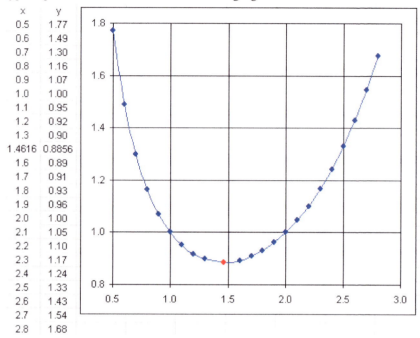

x	y
0.5	1.77
0.6	1.49
0.7	1.30
0.8	1.16
0.9	1.07
1.0	1.00
1.1	0.95
1.2	0.92
1.3	0.90
1.4616	0.8856
1.6	0.89
1.7	0.91
1.8	0.93
1.9	0.96
2.0	1.00
2.1	1.05
2.2	1.10
2.3	1.17
2.4	1.24
2.5	1.33
2.6	1.43
2.7	1.54
2.8	1.68

The algorithm is simple to implement:

```
double Brent(double x1,double x3,double
   func(double),int itmax,double eps,double*x2)
  {
  int iter;
  double x,y1,y2,y3;
  *x2=(x1+x3)/2.;
  y1=func(x1);
  y2=func(*x2);
  y3=func(x3);
  if(y2>=y1||y2>=y3)
    return(y2);
  for(iter=0;iter<itmax;iter++)
    {
    x=y1*(*x2-x3)+y2*(x3-x1)+(x1-*x2)*y3;
    if(fabs(x)<eps*(x3-*x2))
       break;
    x=((*x2*(*x2)-x3*x3)*y1+(x3*x3-x1*x1)*y2+(x1*x1-
    *x2*(*x2))*y3)/x/2.;
    if(x<*x2-eps)
       {
```

```
      x3=*x2;
      y3=y2;
      }
    else if(x>*x2+eps)
      {
      x1=*x2;
      y1=y2;
      }
    else
      break;
    *x2=x;
    y2=func(*x2);
    }
  return(y2);
  }
```

This example converges in 5 iterations requiring only 8 function calls. The program and spreadsheet can be found in the examples\Brent folder. We will use this same algorithm for step length optimization in subsequent multi-variable problems.

Chapter 2. Multi-Valued Functions of One Variable

We will next consider problems, which have more than one solution; specifically, functions of one variable that have more than one zero. This is basically the same as the first problem, only broken up into intervals.

Bessel Function Zeroes

We begin by finding the zeroes of Bessel functions. Why? Because Bessel functions come with Excel® and finding the zeroes of cos(x) or sin(x) seems pretty silly. The first 4 Bessel functions of the second kind (J_0, J_1, J_2, and J_3) are:

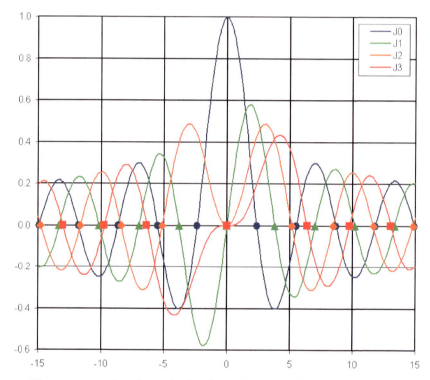

The roots are also shown in the figure by spots of the same color as the curves. These are easily found using a bisection search so that the derivative is not required. The simplest way to accomplish this is through a re-entrant subroutine (i.e., a function that calls itself). We start with the interval -15 to +15 and keep splitting it in half until each interval contains a single root, then we find that root using a bisection search.

```
Sub FindRoot(n As Integer, ByVal x1 As Double, ByVal x2
    As Double, l As Integer)
  If (l > 0) Then
    Call FindRoot(n, x1, (x1 + x2) / 2, l - 1)
```

```
    Call FindRoot(n, (x1 + x2) / 2, x2, 1 - 1)
    GoTo the_end
  End If
  Dim i As Integer, b1 As Double, b2 As Double, x As
    Double
  b1 = Application.Run("BESSELJ", x1, n)
  b2 = Application.Run("BESSELJ", x2, n)
  If (b1 * b2 > 0#) Then GoTo the_end
  For i = 1 To 32
    x = (x1 + x2) / 2
    If (Application.Run("BESSELJ", x, n) < 0) Then
      If (b1 < b2) Then
        x1 = x
      Else
        x2 = x
      End If
    ElseIf (b1 < b2) Then
      x2 = x
    Else
      x1 = x
    End If
  Next i
  For i = 2 To 11
    If (IsEmpty(Cells(i, n + 7))) Then
      Cells(i, n + 7).Value = x
      Exit For
    End If
  Next i
the_end:
End Sub
Private Sub CommandButton1_Click()
  Range("G2:J12").Select
  Selection.ClearContents
  Dim n As Integer
  For n = 0 To 3
    Call FindRoot(n, -15, 15, 4)
  Next n
End Sub
```

Notice that you must specify *byVal* (i.e., by *value* in contrast to by *reference*) in the function declaration so that the value of x1 and x2 are only changed locally. Note also that re-entrant code must always have a mechanism to end the cycle; otherwise, it will get deeper and deeper until you run out of memory—typically resulting in a stack overflow. The files can be found in the online archive in the folder examples\Bessel.

Legendre Polynomial Roots

We will next consider a very useful root finding procedure and one that employs the Newton-Raphson Method: Legendre polynomials. The reason this is useful is that the roots are the abscissas for Gauss Quadrature, the premier method of numerical integration. Graphically, this problem is quite similar to the previous one, as illustrated in this next figure:

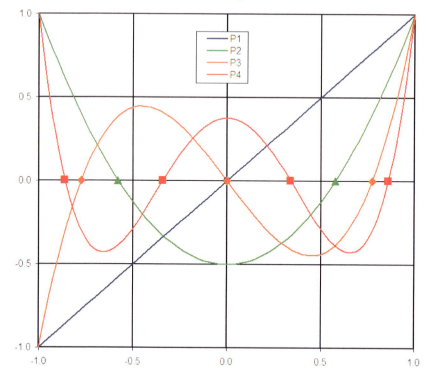

The same procedure with minor modifications will suffice and is included in the Excel® spreadsheet Legendre.xls, which can be found in the online archive in the examples\Legendre folder, along with the complete C code to find any number of roots more accurately using the Newton-Raphson Method. The function calculates the polynomial and its derivative to facilitate this process in a minimal number of steps:

```
void Legendre(int n,double X,double*P3,double*Q3)
  {
  int i;
  double P1,P2,Q1,Q2;
  *P3=1;
  *Q3=0;
  if(n<1)
    return;
  P2=*P3;
  Q2=*Q3;
  *P3=X;
  *Q3=1;
  if(n<2)
```

```
      return;
   for(i=2;i<=n;i++)
      {
      P1=P2;
      Q1=Q2;
      P2=*P3;
      Q2=*Q3;
      *P3=2*X*P2-P1-(X*P2-P1)/i;
      *Q3=Q1+(2*i-1)*P2;
      }
   }
```

The roots are estimated, refined, and inserted into a list in ascending order by the following code:

```
   double RefineRoot(int n,double X1,double X2)
      {
      int i;
      double dP,P,X,Xo;
      X=(X1+X2)/2;
      for(i=0;i<20;i++)
         {
         Legendre(n,X,&P,&dP);
         if(fabs(dP)<0.5)
            break;
         Xo=X;
         X=max(X1,min(X2,X-P/dP));
         if(fabs(P)<=tiny||fabs(X-Xo)<=tiny)
            return(X);
         }
      return(2);
      }

   int InsertRoot(double*Roots,int nr,double Root)
      {
      int i,j;
      for(i=0;i<nr;i++)
         if(fabs(Roots[i]-Root)<tiny)
            return(0);
      for(i=1;i<nr;i++)
         if(Roots[i-1]<Root&&Root<Roots[i])
            break;
      for(j=nr;j>i;j--)
         Roots[j]=Roots[j-1];
      Roots[i]=Root;
      return(i);
      }

   int FindRoots(int n,double*Roots,double*Weights)
      {
      int i,j,k,nr;
```

```c
  double dP,dX,P,Root,X,X1,X2;
  printf("computing coefficients\n");
  if(n<2)
    return(0);
  nr=1;
  if(n%2)
    Roots[0]=0;
  else
    {
    Roots[0]=RefineRoot(n,0.,1.3/pow(n,0.94));
    if(n<3)
      goto weights;
    }
  Roots[nr++]=RefineRoot(n,1.-2.1/pow(n,1.9),1);
  if(n<5)
    goto weights;
  printf("\r%i roots found out of %i",nr,(n+1)/2);
  while(nr<(n+1)/2)
    {
    for(i=1;i<nr;i++)
       {
       k=(n+1)/2-nr+1;
       dX=(Roots[i]-Roots[i-1])/k;
       X2=Roots[i-1];
       for(j=0;j<k;j++)
         {
         X1=X2;
         X2+=dX;
         Root=RefineRoot(n,X1,X2);
         if(Root>1.5)
            continue;
         if(!InsertRoot(Roots,nr,Root))
            continue;
         nr++;
         printf("\r%i roots found out of %i",nr,(n+1)/2);
         break;
         }
       if(nr>=(n+1)/2)
         goto weights;
       }
    }
 weights:
  if(n>4)
    printf("\n");
  for(i=0;i<(n+1)/2;i++)
    {
    X=Roots[i];
    Legendre(n,X,&P,&dP);
    Weights[i]=2/(1-X*X)/dP/dP;
```

```
    }
return((n+1)/2);
}
```

Chapter 3. Functions of Several Variables

We will next consider problems, which have more than one variable. Most often, the desired solution is a global minimum or maximum. Minima and/or maxima are called extrema (singular: extremum). This process might be illustrated by finding the bottom of the following surface:

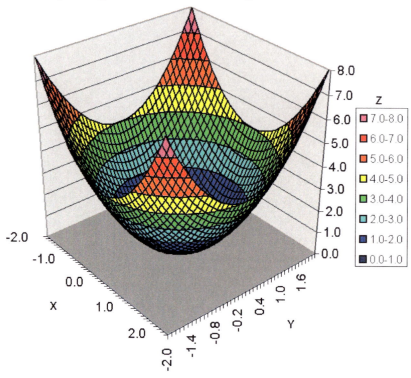

This surface is defined by $z=x^2+y^2$. The low occurs at $x=y=0$ where $\partial z/\partial x = \partial z/\partial y = 0$. Extending Equation 1.1 to multiple dimensions, we arrive at:

$$\begin{bmatrix} x_{k+1} \\ y_{k+1} \end{bmatrix} = \begin{bmatrix} x_k \\ y_k \end{bmatrix} - \frac{1}{damp} \begin{bmatrix} \dfrac{\partial^2 z}{\partial x^2} & \dfrac{\partial^2 z}{\partial x \partial y} \\ \dfrac{\partial^2 z}{\partial x \partial y} & \dfrac{\partial^2 z}{\partial y^2} \end{bmatrix}^{-1} \begin{bmatrix} \dfrac{\partial z}{\partial x} \\ \dfrac{\partial z}{\partial y} \end{bmatrix} \qquad (3.1)$$

The square matrix containing the second partial derivatives (or it's inverse) is called a Hessian. For this first problem (first tab in bump.xls) the partial derivatives are easily calculated analytically:

$$\frac{\partial z}{\partial x} = 2x$$

$$\frac{\partial z}{\partial y} = 2y$$

$$\frac{\partial^2 z}{\partial x^2} = \frac{\partial^2 z}{\partial y^2} = 2 \qquad (3.2)$$

$$\frac{\partial^2 z}{\partial x \partial y} = 0$$

The iteration described in Equation 3.1 can be programmed in Excel® using the MMINVERSE(), MMULT(), and TRANSPOSE() functions. In this case, it looks like:

x	y	dz/dx	dz/dy	d²z		step
		starting at (0.5,0.5)				
0.50	0.50	1.00	1.00	2.00	0.00	0.500
				0.00	2.00	0.500
0.00	0.00	0.00	0.00			
		starting at (-0.5,0.5)				
-0.50	0.50	-1.00	1.00	2.00	0.00	-0.500
				0.00	2.00	0.500
0.00	0.00	0.00	0.00			
		starting at (0.5,-0.5)				
0.50	-0.50	1.00	-1.00	2.00	0.00	0.500
				0.00	2.00	-0.500
0.00	0.00	0.00	0.00			
		starting at (-0.5,-0.5)				
-0.50	-0.50	-1.00	-1.00	2.00	0.00	-0.500
				0.00	2.00	-0.500
0.00	0.00	0.00	0.00			

Because the first derivatives are linear and the second are constant, the process converges in a single step, regardless of the starting value. Such is not the case with any meaningful problem. We will next consider the function $z=9x/exp(x^2+y^2)$, which has both a minimum and a maximum. The iteration for this problem also converges quickly, as long as a fortuitous initial guess is provided. Whether the iterations end up at the low or high spot depends on the starting value, much as would be the case if you placed a marble on the surface and let it roll into the valley or released a little balloon beneath the surface and let it rise up into the peak. The surface looks like:

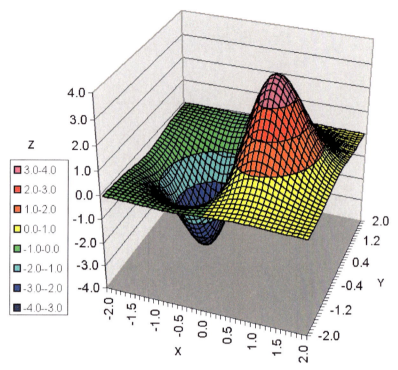

Starting at x=0.5, y=0.5, the iterations proceed as follows:

\multicolumn{7}{c}{start at (0.5,0.5) looking for maximum}						
x	y	dz/dx	dz/dy	d²z		step
0.50	0.50	2.73	-2.73	-13.65	-2.73	-0.50
	z=	2.73	-2.73	-2.73	-2.73	1.50
0.75	-0.25	-0.60	1.81	-13.55	-0.30	0.05
	z=	3.61	-0.30	-6.32		-0.29
0.72	-0.11	-0.26	0.81	-14.88	-0.06	0.02
	z=	3.81	-0.06	-7.46		-0.11
0.72	-0.05	-0.13	0.40	-15.21	-0.01	0.01
	z=	3.85	-0.01	-7.66		-0.05
0.71	-0.03	-0.06	0.20	-15.34	0.00	0.00
	z=	3.86	0.00	-7.70		-0.03
0.71	-0.01	-0.03	0.10	-15.39	0.00	0.00
	z=	3.86	0.00	-7.72		-0.01
0.71	-0.01	-0.02	0.05	-15.42	0.00	0.00
	z=	3.86	0.00	-7.72		-0.01
0.71	0.00	-0.01	0.02	-15.43	0.00	0.00
	z=	3.86	0.00	-7.72		0.00
0.71	0.00	0.00	0.01	-15.43	0.00	0.00
	z=	3.86	0.00	-7.72		0.00

After 8 steps, the solution arrives at x=0.71, y=0.0, z=3.86, which is the top of the bump. The Hessian is outlined in red, the right side of Equation 3.1 is outlined in blue, and the left side (i.e., the next step) is outlined in green. If the iteration begins closer to the pit, it will converge to the minimum.

start at (-0.5,-0.5) looking for minimum

x	y	dz/dx	dz/dy	d²z		step
-0.50	-0.50	2.73	-2.73	13.65	2.73	0.50
		z=	-2.73	2.73	2.73	-1.50
-0.75	0.25	-0.60	1.81	13.55	0.30	-0.05
		z=	-3.61	0.30	6.32	0.29
-0.72	0.11	-0.26	0.81	14.88	0.06	-0.02
		z=	-3.81	0.06	7.46	0.11
-0.72	0.05	-0.13	0.40	15.21	0.01	-0.01
		z=	-3.85	0.01	7.66	0.05
-0.71	0.03	-0.06	0.20	15.34	0.00	0.00
		z=	-3.86	0.00	7.70	0.03
-0.71	0.01	-0.03	0.10	15.39	0.00	0.00
		z=	-3.86	0.00	7.72	0.01
-0.71	0.01	-0.02	0.05	15.42	0.00	0.00
		z=	-3.86	0.00	7.72	0.01
-0.71	0.00	-0.01	0.02	15.43	0.00	0.00
		z=	-3.86	0.00	7.72	0.00
-0.71	0.00	0.00	0.01	15.43	0.00	0.00
		z=	-3.86	0.00	7.72	0.00

After 8 steps, the solution arrives at x=-0.71, y=0, z=-3.86, the bottom of the pit. Following the direction defined by the inverse of the Hessian matrix times the first derivatives has often been called the *Method of Steepest Descent* (or *Gradient Descent*). These two examples illustrate this principle quite well. There are two problems with this method, which we will now discuss: 1) the method will sometimes overshoot (as in the one-dimensional cases already presented) and 2) more often than not, we don't have neat analytical expressions for the derivatives.

The problem with step length and starting values is much worse that this illustration would seem to indicate. In fact, I chose the starting values (-0.5,-0.5) and damping factor (*damp* in Equation 3.1 equals 2) so that a solution would be found. There are countless other choices that—rather than improving—get worse with each iteration. Consider the case of starting at x=y=0.6 with no damping (*damp*=1). The first step hops over the solution and almost out of the graph altogether. The second and subsequent steps are worse still. These are illustrated in this next figure:

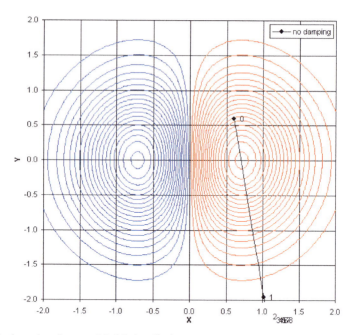

A damping factor of 1.6 is hardly better:

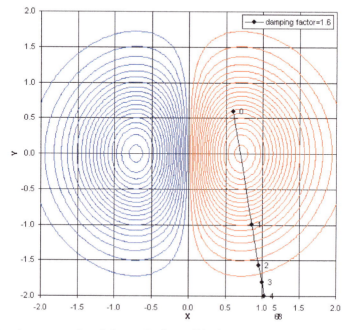

Starting at x=y=1 and *damp*=5 takes off in the wrong direction:

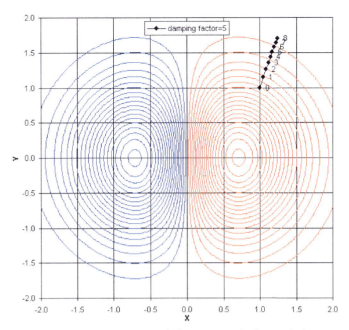

At least starting with x=y=0.5 and *damp*=5 results in a solution.

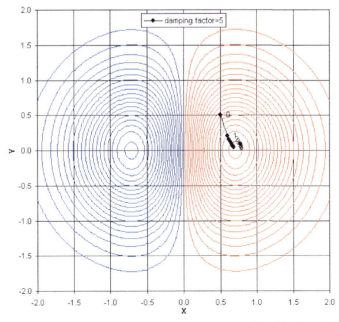

If the method doesn't work reliably or even predictably for this simple case, how can we solve practical problems? The third example in this last set of four shows that the iterations may even take off in the wrong direction. Heavily damping (i.e., a large damping factor) and lots of small iterations is neither efficient nor sufficient in many cases. The figure below shows the starting values (initial guess) that ultimately converge using Newton's Method with a damping factor of 5 for the preceding problem.

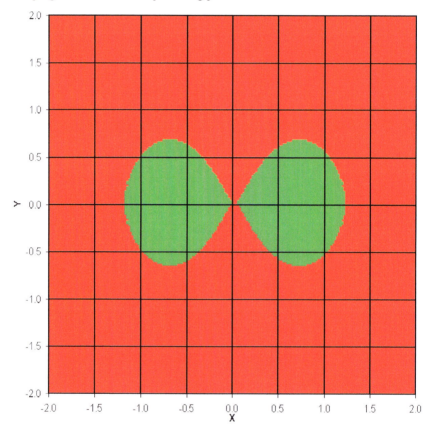

The green regions are sometimes called *attraction* zones. In this case, these zones are quite small, considering the range of possible values. Even within these zones it takes on average 32 iterations to converge. It is no wonder that considerable effort has been devoted to solving these types of problems and various methods abound.

Newton methods take the following form:

$$\begin{bmatrix} next \\ step \end{bmatrix} = \begin{bmatrix} this \\ step \end{bmatrix} - \alpha \begin{bmatrix} Hessian \\ of \\ function \end{bmatrix} \begin{bmatrix} gradient \\ of \\ function \end{bmatrix} \qquad (3.3)$$

where α is the step length parameter. We don't always have the second partial derivatives, although the first derivatives are often available; thus, the gradient can be easily calculated, but the Hessian cannot. As the preceding examples illustrate, the simple Newton Method may be inadequate.

Recall that the iterations in the third case went off in the opposite direction of the minimum? If the solution doesn't improve, why not try reversing the step? In the first two cases, the iterations overshot the solution, so why not try 1/2, 1/4, 1/8, etc. A simple modification that often greatly improves convergence is the following sequence of damping factors: 1, -1, 2, -2, 4, -4, 8, -8, …, which is produced by the following code for two variables, which is easily extended to an arbitrary dimension:

```
r1=residual(x1,y1);
damp=1.;
for(j=0;j<16;j++)
  {
  x2=x1-drdx/damp;
  y2=y1-drdy/damp;
  r2=residual(x2,y2);
  if((increasing&&r2>r1)||((!increasing)&&r2<r1))
    break;
  if(j%2)
   damp=-damp*2.;
  else
    damp=-damp;
  }
```

Chapter 4: Bounding & Scaling

More often than not, valid solutions are limited to some range of values. It is rare that any value (i.e., $-\infty < x < \infty$) will do for a solution. Many of the methods proposed for solving nonlinear problems ignore this important fact. In most cases, this easily handled by the following transformation:

$$x = x_{min} + (x_{max} - x_{min})\left[\frac{1 + \tanh(x')}{2}\right] \quad (4.1)$$

Regardless of the value of x', x will be limited to $x_{min} < x < x_{max}$. Near the center, $(x_{max} + x_{min})/2$, the relationship is approximately linear. The hyperbolic tangent function is available in Excel® as well as most programming languages, including C. Most of the range is included between -2 and +2, so that the figures in the preceding chapter serve as useful illustrations of this domain. Another transformation often suggested is:

$$x' = \frac{ax}{(x - x_{min})(x_{max} - x)} \quad (4.2)$$

both are shown in the following figure:

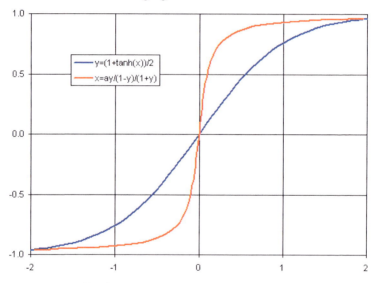

For most applications, Equation 4.2 is much too steep and results in unwanted oscillations. It is also more difficult to implement, involving branches and the square root. Some have suggested using $sin(x)$ or $cos(x)$ instead of $tanh(x)$ in Equation 4.1, but this is not a good choice. It is best to stick with something that has one and only one value of x' for any value of x.

Scaling can be important too. The simplest scaling would be to divide the residual by the span (maximum minus minimum values) or the sum of the squares of the target values. It is rare that a nonlinear problem has an exact solution (i.e., one resulting in a zero residual or precise agreement with the objective). More often than not, nonlinear problems consist of minimizing or maximizing some residual. Many algorithms are designed to either minimize or maximize, but not both. As long as the residual is never zero, the reciprocal of one is the other.

Most any method will work best when a map of the residual looks like a salad bowl, such as the first figure in Chapter 3. Drop a marble anywhere inside a salad bowl and it will eventually find the bottom. Any method worth publishing will converge under the right conditions. It's equally true that most methods will diverge under the wrong conditions. Regardless of how many dimensions (i.e., independent variables) you're working with, it's wise to formulae the problem statement so that the residual approximates this shape.

Residuals may vary over orders of magnitude. When this is the case, you will likely have trouble finding a suitable solution. As the residual is rarely zero, taking the log of the residual may greatly improve the search. Taking the log may also complicate refinement of the solution, as slight variations in the residual may disappear. When this happens, use the log to find an approximate solution, then refine the solution without the log.

Chapter 5: Nonlinear Least-Squares

One of the most common tasks requiring solution of nonlinear equations is that of curve fitting when the function (or functions) being fit to the data are not linear. Sometimes this arises from the shape of the data and sometimes it arises from knowing the analytical solution but not the properties. For example, data, which exhibits one or more asymptotes (i.e., goes straight up or down) or levels off to some constant value. Polynomials (i.e., $y=a+bx+cx^2+...$) will never exhibit this behavior, regardless of how many terms you add. Data that goes straight up or down may sometimes be approximated by: $y=a/(x-b)$. Asymptotic behavior may sometimes be approximated by: $y=(a+bx)/(1+cx)$. Solving this problem so as to minimize the sum of the squares of the residuals (i.e., the difference between the approximation and the data) is called *nonlinear least squares*.

The first problem we will consider here is one, which arises from a known analytical solution. The response begins at $t=f=0$, rises to a maximum, and then falls off asymptotically. The functional shape is $f(t)=at/e^{-t/\tau}$. The data and approximation are shown in the following figure:

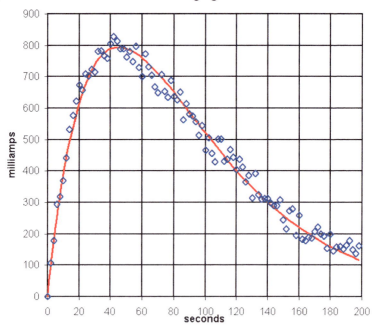

The parameters a and τ are easily calculated with Excel® using the Solver Add-In. Simply guess initial values, create a column of calculated values, then a column of error, and finally set a cell to the sum of the squares of the errors. The Solver will adjust the two (or more) parameters to minimize SUMSQ().

The result yielding the red curve is: a=47.8, τ=45.2. This example can be found in the online archive in folder examples\nllsq nonlinear_least_squares.xls.

The Excel® Solver Add-In is a remarkable tool, which uses several algorithms to obtain a solution. Often this is adequate, but we want to explore how this works, why it sometimes doesn't work, and how to automate and customize the technique in the event we need to process data in production mode or inside a larger software application. The log of the residual for this problem is shown below:

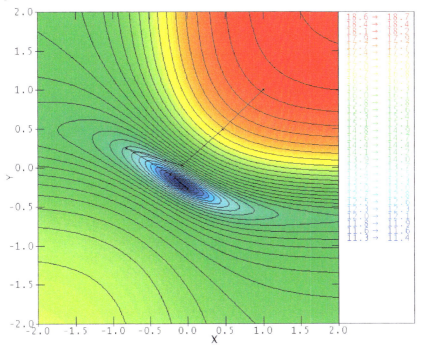

The minimum and maximum values of both a and τ are 25 and 75, respectively. The transform coordinates x and y are used with Equation 4.1 to span these values so that the iterations can't possibly get too far out of bounds. The optimum solution is at the center of the innermost blue ring. Newton's Method (presented in Chapter 3 along with the modification at the end of the chapter) will quickly find a solution, as illustrated by the segmented path above. The code can be found in nllsq.c.

A similar problem also involves inferring properties from experimental data. In this case, the response is linear at first, rises to a maximum, and levels off at that value. The ultimate value is easily calculated by average the final portion of the data. Determining the time constant, τ, and shape factor, a, is a problem in nonlinear minimization. The analytical function that best exhibits this behavior is the hyperbolic tangent, $y=c*tanh((t/tau)^a)$.

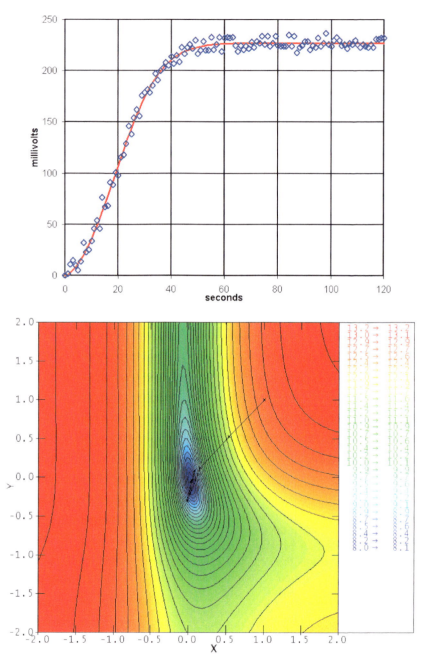

This problem can also easily be solved using the Excel® Solver as before. Supply initial guesses, create a column of calculated values, a column of errors, and a cell containing SUMSQ(). The asymptotic value, c=227, the time constant, τ=29.8, and the exponent is 2/3. The data and solution are shown in the preceding figure. It is not necessary (or helpful) to take the log of the residual in this case. The residual and solution path is mapped below in the transform coordinates.

In both of these examples, the gradient and Hessian are calculated using finite differences. This is often inconvenient, slow, and inaccurate. Such is often the case, providing more than adequate motivation for developing what are called *Quasi-Newton Methods*, that is, methods that are roughly based on Newton's Method but do not require the gradient and/or Hessian.

Chapter 6: Quasi-Newton Methods

The need to solve practical problems has given rise to what are called *Quasi-Newton Methods*, that is, methods that are roughly based on Newton's Method, but have been modified to address the lack of convenient analytical expressions for the partial derivatives.

Davidon-Fletcher-Powell Method

DFP was the first technique to extend the Secant Method to multi-dimensions.[6] DFP is a rank-two method that uses the following update formula for the inverse of Hessian matrix, $B = H^{-1}$:

$$B_{k+1} = B_k + \gamma\, s_k s_k^T + \beta\, p_k p_k^T \quad (6.1)$$

$$s_k = x_{k+1} - x_k \quad (6.2)$$

$$y_k = \nabla f(x_{k+1}) - \nabla f(x_k) \quad (6.3)$$

$$p_k = B_k y_k \quad (6.4)$$

$$\gamma = \frac{1}{y_k^T s_k} \quad (6.5)$$

$$\beta = -\frac{1}{y_k^T p_k} \quad (6.6)$$

Scalars γ and β plus the vector p are selected so as to follow the Secant Method and achieve second order agreement with a Taylor series expansion of the function being approximated. This iterative update also assures that the Hessian is symmetric and positive definite. The direction vector, d_k, arises from Equation 3.1, where Δ_k is the gradient at x_k:

$$d_k = -B_k \nabla f(x_k) \quad (6.7)$$

The iteration typically begins with B equal to the identity matrix. As we have already seen, assuming a positive step length, $\alpha > 0$, may lead away from the solution. It may also take quite a few steps to find the value of α that minimizes (or maximizes) the function (i.e., minimizes the gradient along the direction d_k.

---

[6] Johnson, I. L., Jr., "The Davidon-Fletcher-Powell Penalty Function Method: A Generalized Iterative Technique for Solving Parameter Optimization Problems," NASA Technical Note D-8251, 1976.

We will first consider the problem in which $z=(cosh(x-0.5)+(y-0.5)^2)/8$, shown in the following figure:

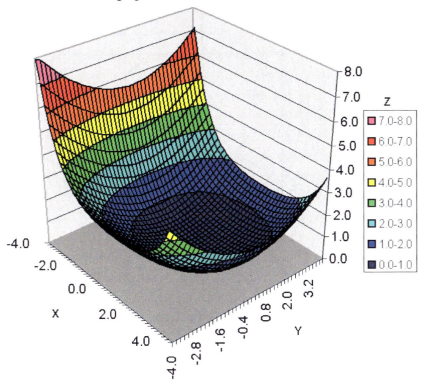

Beginning with a guess of $x=y=-3$, we apply the DFP Method along with Brent's Method to locate the minimum, which is at $x=y=0.5$. The source code (dfpmin.c), equations, and accompanying spreadsheet can be found in the examples\dfp folder. The solution follows the path shown in the following figure. This is a very simple problem and the partial derivatives are easily calculated, which is rarely the case with practical applications.

```
double func(double*p)
  {
  calls.func++;
  return((cosh(p[0]-0.5)+sq(p[1]-0.5))/8.);
  }
void grad(double*p,double*g)
  {
  calls.grad++;
  g[0]=sinh(p[0]-0.5)/8.;
  g[1]=(2.*p[1]-1.)/8.;
  }
```

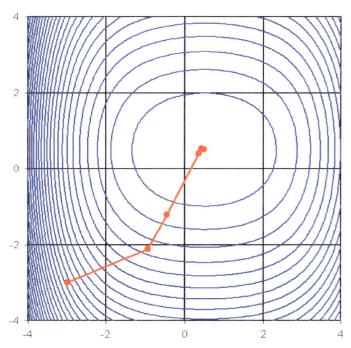

Initializing the Hessian to the identity matrix is often a poor choice. It works in this case because the residual is shaped like a salad bowl and the magnitude of the residual sort of fits with the dimensions of the solution domain. The algorithm itself is quite simple:

```
f2=func(p);
grad(p,g);
for(i=0;i<n;i++)
   {
   for(j=0;j<n;j++)
     h[n*i+j]=0.;
   h[n*i+i]=1.;
   d[i]=-g[i];
   }
for(it=1;it<=itmax;it++)
   {
   a=1.;
   for(jt=0;jt<32;jt++)
      {
      for(i=0;i<n;i++)
        s[i]=p[i]+d[i]/a;
      f2=func(s);
      if(f2<*f)
        break;
      if(jt==0)
```

```
       a=2.;
    else if(jt%2)
       a=-a;
    else
       a=-a*2.;
    }
  for(i=0;i<n;i++)
    {
    s[i]=d[i]/a;
    p[i]+=s[i];
    }
  f1=*f;
  *f=f2;
  if(2.*fabs(f1-f2)<=epsilon*(fabs(f1)
+fabs(f2)+epsilon))
     break;
  for(i=0;i<n;i++)
     y[i]=g[i];
  grad(p,g);
  for(i=0;i<n;i++)
     y[i]=g[i]-y[i];
  for(i=0;i<n;i++)
     for(q[i]=j=0;j<n;j++)
        q[i]+=h[n*i+j]*y[j];
  for(yTs=yHy=i=0;i<n;i++)
     {
     yTs+=y[i]*s[i];
     yHy+=y[i]*q[i];
     }
  for(i=0;i<n;i++)
     for(j=0;j<n;j++)
        h[n*i+j]+=-q[i]*q[j]/yHy+s[i]*s[j]/yTs;
  for(i=0;i<n;i++)
     for(d[i]=j=0;j<n;j++)
        d[i]-=h[n*i+j]*g[j];
  }
```

We can also use the DFP Method to solve the two problems from the end of the last chapter. The DFP method does find the correct solution to the first problem, but arrives there through a circuitous path, as shown in the following figure. The gradient is calculated as before by finite difference, but that's not the problem. The poor initial estimate of the Hessian is.

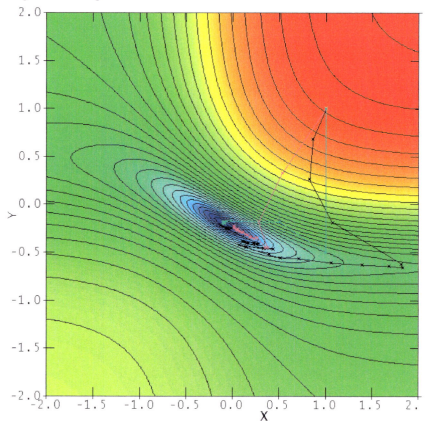

The DFP solution path is shown in black. The magenta line is the solution path for the next method presented below and the dark cyan line is the solution path for the method presented in the next chapter.

A solution is found for the second by a much more direct path, well approximating the steepest descent or Newton's Method. This is purely coincidental.

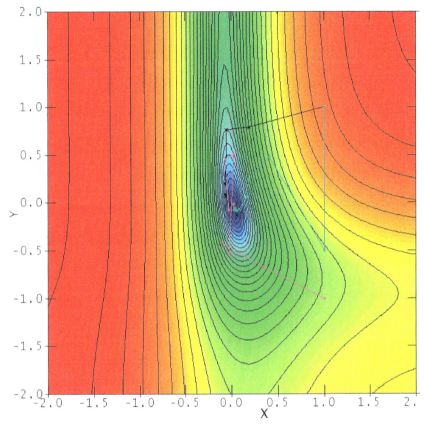

The DFP solution path is shown in black. The magenta line is the solution path for the BGFS method presented below and the dark cyan line is the solution path for Broyden's Method, which is presented in the next chapter.

Broyden-Fletcher-Goldfarb-Shanno Method

The BFGS Method was introduced after the DFP Method and is often said to be preferable to it.[7,8] In practice which method is faster and/or more accurate depends on the problem, as illustrated in the two preceding figures. The BFGS uses a different Hessian update formula:

$$H_{k+1} = H_k + \gamma \, y_k y_k^T + \beta \, p_k p_k^T \tag{6.8}$$

$$p_k = H_k s_k \tag{6.9}$$

$$\gamma = \frac{1}{y_k^T s_k} \tag{6.10}$$

$$\beta = -\frac{1}{s_k^T p_k} \tag{6.11}$$

The search direction is given by:

$$H_k d_k = -\nabla f(x_k) \tag{6.12}$$

A variant of the BFGS algorithm that generates the inverse Hessian ($B=H^{-1}$) with each step is given by the following:

$$B_{k+1} = B_k - \gamma \left(q_k s_k^T + s_k q_k^T \right) + \varphi \, s_k s_k^T \tag{6.13}$$

$$q_k = B_k y_k \tag{6.14}$$

$$\phi = \frac{s_k^T y_k + q_k^T y_k}{\left(s_k^T y_k \right)^2} \tag{6.15}$$

Calculating the gradient with finite differences at each step often takes much longer than inverting a small matrix, so that eliminating the step of solving a small set of linear equations (i.e., inverting H to get B) is of little consequence and may take equally as long as Equation 6.13 over Equation 6.8. The BFGS solution path is shown in magenta in the two preceding figures.

---

[7] Goldfarb, D., "A Family of Variable Metric Methods Derived by Variational Means," Mathematics of Computation Journal, Vol. 24, pp. 23–26, 1970.
[8] Shanno, D. F., "Conditioning of Quasi-Newton Methods for Function Minimization," Mathematics of Computation Journal, Vol. 24 pp. 647–657, 1970.

An exploded view in the vicinity of the solution is shown below:

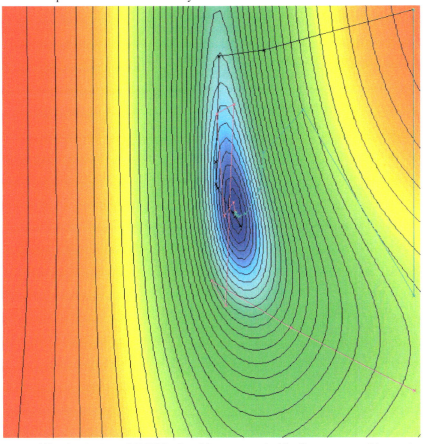

While the DFP and BFGS algorithms are suitable for many problems, it would be *really* convenient to have a method that didn't require derivatives!

Chapter 7. Broyden's Derivative-Free Algorithm

Broyden—one of the developers of the BFGS Method—also devised a derivative-free method.[9] This technique is particularly attractive, as the gradient and Hessian are rarely available for practical problems. In the next chapter we will consider another way of dealing with this problem. We will first consider Broyden's method, which doesn't require derivatives. This method uses all the residuals (instead of just the sum of the squares), which often greatly exceeds the number of independent parameters, especially in regression problems.

Rather than starting with Newton's Method and updating an estimate of the Hessian, Broyden's begins with the *Jacobian*. The elements of the Jacobian are given by:

$$J_{i,j} = \frac{\partial r_i}{x_j} \tag{7.1}$$

where r_i is the residual term *i*. For a least-squares problem, the terms of the Hessian are:

$$H_{i,j} = \frac{\partial^2 (r^T r)}{\partial x_i \partial x_j} \tag{7.2}$$

The Hessian is always square (*nxn*); whereas, the Jacobian is often rectangular (*mxn*, *m>n*). It also stands to reason that we retain more information about the residual in the Jacobian than the Hessian, as it has a greater number of rows. Broyden's method in terms of the same variables as before can be expressed by:

$$J_k s_k = y_k \tag{7.3}$$

$$J_{k+1} = J_k + \frac{(y_k - J_k s_k) s_k^T}{s_k^T s_k} \tag{7.4}$$

There is also an inverse update to the Broyden method, but this is often problematic and not worth the savings in solving a few simultaneous linear equations.[10]

$$J_{k+1}^{-1} = J_k^{-1} + \frac{(s_k - J_k^{-1} y_k) y_k^T}{y_k^T y_k} \tag{7.5}$$

The most convenient aspect of Broyden's method is that initialization of the Jacobian is so easily accomplished. Simply hop around the domain a bit and

---

[9] Broyden, C., "A New Method of Solving Nonlinear Simultaneous Equations," *Computational Journal*, Vol. 12, pp. 94-99, 1969.

[10] Kvaalen, E., "A Faster Broyden Method," SIAM BIT Numerical Mathematics, Vol. 31, No. 2, pp. 369–372, November, 1991.

update Equation 7.4. This is a rank-one update and information is only added in the space traversed, eventually approximating the true Jacobian. As for the performance, the Broyden Method solution path has already been illustrated in the two problems of the previous chapter (the dark cyan lines). In both of these examples, the Broyden path is more direct than the DFP, but not quite as direct as the BFGS. In both cases, it takes fewer steps to reach this solution.

I first published a paper on Broyden's Method in 1991, which included some enhancements, including escape from extraneous entrapment in local minima. The text of this paper is included in the final appendix. All of the associated software can be found in the online archive in the folder examples\Broyden. When trapped, simply hop around a bit and it will often escape and find a better solution.

The next problem we will solve using Broyden's Method to is $r_1=75*x_1^2-169*x_2^2/9$, $r_2=845*x_1*x_2^2/3-125*x_1^3-1$. The solution path is shown below:

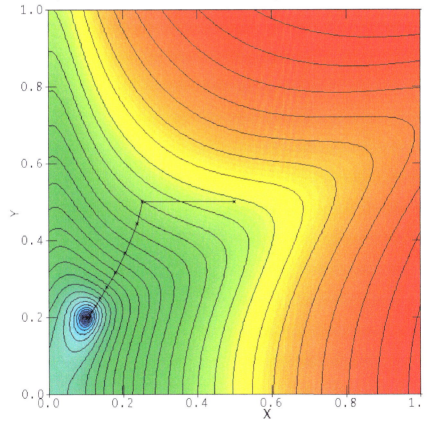

Of course, many problems have more than two unknowns and can't be illustrated in two-dimensional space. The following problem has three:

$f_1=15*x_1-cos(5/9*(x_2-1/5)*\pi*x_3)-1/2$, $f_2=25*x_1{}^2-81*(x_2-1/10)^2-sin(5*\pi*x_3/9)+1$, $f_3=exp(-5*x_1*(x_2-1/5))-100*\pi*x_3/9+10*\pi/3-1$. One way of presenting the solution path is a shaded cube as seen from three sides:

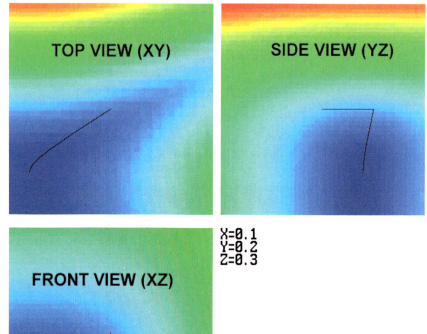

The solution path starts in the middle ($x_1=x_2=x_3=0.5$) and proceeds to the center of the blue spot, near $x_1=0.1$, $x_2=0.2$, $x_3=0.3$.

A similar problem is described by the equations: $f_1 = x_1 + \cos(5*(x_1-1/10)*x_2*x_3/3) - 11/10$, $f_2 = |x_1 - 11/10|^{1/4} + x_2/2 + 5*x_3^2/9 - x_3/2 - 1$, $f_3 = -(x_1-1/10)^2 - x_2^2/40 + x_2/200 + 10*x_3/3 - 1$.

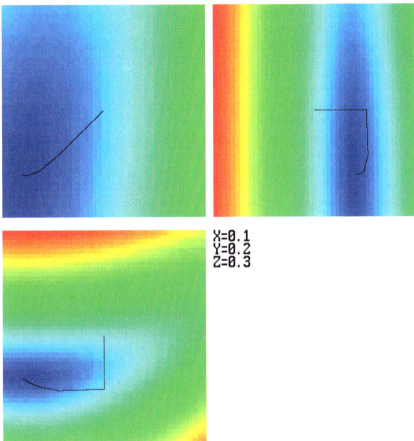

This solution path also starts in the middle ($x_1=x_2=x_3=0.5$) and proceeds to the center of the blue spot, near $x_1=0.1$, $x_2=0.2$, $x_3=0.3$.

Chapter 8. Approximating Gradients and Hessians

In the preceding examples we have already seen the most common way of approximating gradients (first partial derivatives) and Hessians (second partial derivatives)—finite differences. In many cases this will suffice, especially when the residual function is an explicit formula, calculated using double-precision floating-point arithmetic. An entire class of problems exists for which this may not be the case—model-tuning parameters. For instance, we may be calibrating a much more complex model to experimental data. Fields in which such problems arise span the range from ecological systems to the stock market. I have used the following method on many occasions and allude to this in the SIAM paper (Appendix L).

Such problems have one or more of several characteristics, including: 1) results may not be available for any arbitrary inputs, 2) the model may be a separate program that can't be launched or called as function, 3) the number of model runs must be minimized due to computational time or limited resources, 4) the convergence level of the model is such that small changes in tuning parameters produce meaninless differentials.

Case 3 in the SIAM paper is an example of this type problem. The objective was tuning a cooling tower model (FACTS) by simultaneously adjusting heat and mass transfer coefficients. In the original implementation, I created an input file, launched the executable cooling tower model, and read the output file—all inside the nonlinear equation solver loop. It was quite cumbersome and very time consuming. I have modified Case 2 in the online archive to simplify and approximate the program-launching step.

Another program I work with a lot is GateCycle®, a thermodynamic cycle modeling tool sold by General Electric®. GC is especially useful for analyzing combined cycle power systems. While this program is quirky and hasn't been updated since 2005[11], it's still far superior to similar programs currently available. It's fast, but achieves this at the expense of several important behaviors, including convergence. Every time you run GC, you get a slightly different result—even with the exact same inputs. From a programming standpoint, this is completely unacceptable. This problematic behavior arises from the fact that it uses the last solution as the starting point for the next. If the previous solution croaked, you're screwed—another unacceptable outcome. Because the solution always contains some level of noise (imprecise convergence), you can't reliably calculate partial derivatives as you might with a more precise and consistent model.

---

[11] There's a newer version of GateCycle (6.x), but it has the worst user interface of any program ever written—besides being slower than a herd of snails stampeding up the side of a salt dome. GC6 is an xml-based programming atrocity and legendary corporate blunder. General Electric® dumped all of the original developers, who have dispersed to the four winds.

Local Approximation of the Residual

The solution to these challenges is to approximate the residual and it's partial derivatives locally, using a database of model runs. This is how I eventually replaced the cumbersome process in Case 3 of the SIAM paper and how I have tuned many GateCycle® models to historical operational data. The first step is to build a table of runs that span the expected range of tuning parameters. We read the table and use the results to build a least squares approximation of the residual at each step. Since we have the gradient and the Hessian, Newton's Method is the obvious choice. For the case of two tuning parameters (x and y), we have:

$$r = a_0 + a_1 x + a_2 y + a_3 x^2 + a_4 xy + a_5 y^2 \qquad (8.1)$$

Gather enough data points from the table around the current value to assure over-determination of the coefficients—nine in the case of two variables. The code, input tables, and results can be found in the online archive in folder examples\tabular. The approximating function is:

```
double ApproximateSurface(double*p,double*g,double*h)
  {
  int i,j,x,xx,y,yy;
  double A[n],AtA[n*n],AtB[n],B;
  for(xx=1;xx<surface.nx-2;xx++)
    if(p[0]<surface.x[xx+1])
      break;
  for(yy=1;yy<surface.ny-2;yy++)
    if(p[1]<surface.y[yy+1])
      break;
  memset(AtA,0,sizeof(AtA));
  memset(AtB,0,sizeof(AtB));
  for(y=yy-1;y<=yy+1;y++)
    {
    for(x=xx-1;x<=xx+1;x++)
      {
      B=surface.z[surface.nx*y+x];
      A[0]=1.;
      A[1]=surface.x[x];
      A[2]=surface.y[y];
      A[3]=A[1]*A[1];
      A[4]=A[1]*A[2];
      A[5]=A[2]*A[2];
      for(i=0;i<n;i++)
        {
        for(j=0;j<n;j++)
          AtA[n*i+j]+=A[i]*A[j];
        AtB[i]+=A[i]*B;
        }
      }
    }
```

```
    Gauss(AtA,AtB,n);
    if(g)
      {
      g[0]=AtB[1]+2.*AtB[3]*p[0]+AtB[4]*p[1];
      g[1]=AtB[2]+AtB[4]*p[0]+2.*AtB[5]*p[1];
      }
    if(h)
      {
      h[0]=2.*AtB[3];
      h[1]=h[2]=AtB[4];
      h[3]=2.*AtB[5];
      }
    return(AtB[0]
          +AtB[1]*p[0]
          +AtB[2]*p[1]
          +AtB[3]*p[0]*p[0]
          +AtB[4]*p[0]*p[1]
          +AtB[5]*p[1]*p[1]);
 }
```

The Newton iteration is the same as presented at the end of Chapter 3. Results for the two examples are shown in the following figures:

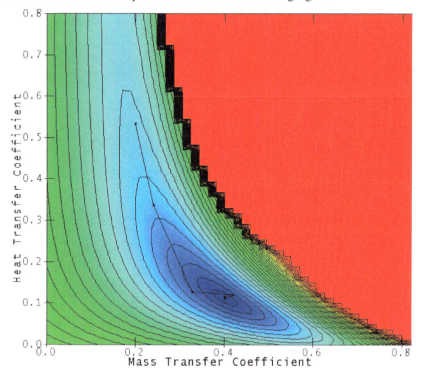

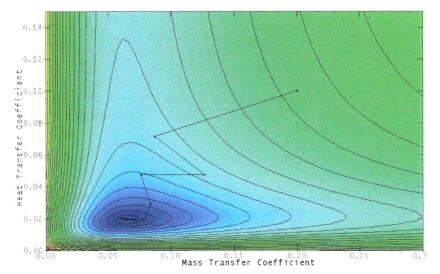

No regression will ever adequately approximate the surfaces illustrated in the preceding two figures. Simply curve-fitting the results and analytically solving for the minimum isn't an option. Merely picking the smallest value in the tabulated results isn't good enough either, considering the coarse resolution.

Broyden's Method from Chapter 7 can also be used with tabulated data, provided the number of residuals is greater than or equal to the number of unknowns. The following example comes directly from a GateCycle® model, in which capacity and heat rate (inversely proportional to thermal efficiency) are calculated based on ambient conditions. The results are shown below:

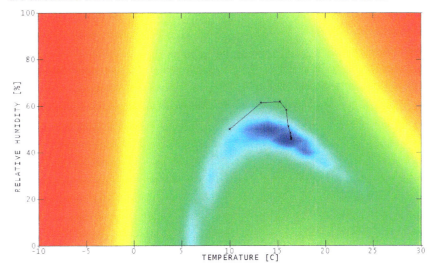

Chapter 9. Multi-Dimensional Bisection Search

In Chapter 1 we considered the number of iterations required to locate a solution and how this may be influenced by the starting location or initial guess. Several of the colored figures revealed large areas where the algorithms failed (Illinois Variant 2 and Anderson-Björck). The last figure in Chapter 3 showed very limited attraction zones (i.e., regions within which an initial guess will result in success of the method, yielding a solution). The first figure in Chapter 8 has a large flat zone (i.e., the whole upper right half of the domain is red). While the gradient vanishes in this zone, it's not the solution we're hoping to find, because it corresponds to a large residual.

Brute Force

The most inefficient way to locate a solution would be to take small steps throughout the entire domain, saving the highest and/or lowest value. For illustration, we will use the second problem from Chapter 2—the "bump" defined by $z=9x/exp(x^2+y^2)$. In order to obtain a solution within $\Delta x=\Delta y=\pm 0.1$, this would require $(4/0.1+1)^2=1681$ function evaluations, as the domain spans -2 to +2 for both x and y. To obtain a solution to within ± 0.01 would require $10^2=100$ times as many function evaluations. If there were three variables instead of two, the former would require 68,921 function evaluations and the latter 69 billion. Almost any intelligent method is more attractive than brute force. Randomly searching a domain makes as much sense as buying lottery tickets. Lotteries are a shameful *Exploitation of the Mathematically Challenged*.

Divide and Conquer

The simplest approach to searching for a solution (or an effective initial estimate from which to further refine a solution) is to divide and conquer. This technique is simple: sequentially subdivide the domain, only refining the search in those subdivisions that yield better (or the best) results. This is easily accomplished with a re-entrant function (i.e., one that calls itself). While this may result in a stack overflow, there's probably more than enough memory on any modern computer to handle it.

The code is simple (example1.c in folder examples\bisection):

```
void bisect(double Xm,double Xx,double Ym,double
   Yx,double func(double,double),int
   n,double*X,double*Y,double*F)
{
double F11,F12,F21,F22,X1,X2,Y1,Y2;
X1=(3.*Xm+Xx)/4.;
X2=(Xm+3.*Xx)/4.;
Y1=(3.*Ym+Yx)/4.;
Y2=(Ym+3.*Yx)/4.;
F11=func(X1,Y1);
F12=func(X1,Y2);
F21=func(X2,Y1);
```

```
F22=func(X2,Y2);
if(F11<=fmin3(F12,F21,F22))
   {
   if(F11<*F)
      {
      *X=X1;
      *Y=Y1;
      *F=F11;
      }
   if(n>0)
      bisect(Xm,(Xm+Xx)/2.,Ym,(Ym+Yx)/2.,func,
  n-1,X,Y,F);
   }
if(F12<=fmin3(F11,F21,F22))
   {
   if(F12<*F)
      {
      *X=X1;
      *Y=Y2;
      *F=F12;
      }
   if(n>0)
      bisect(Xm,(Xm+Xx)/2.,(Ym+Yx)/2.,Yx,func,
  n-1,X,Y,F);
   }
if(F21<=fmin3(F11,F12,F22))
   {
   if(F21<*F)
      {
      *X=X2;
      *Y=Y1;
      *F=F21;
      }
   if(n>0)
      bisect((Xm+Xx)/2.,Xx,Ym,(Ym+Yx)/2.,func,
  n-1,X,Y,F);
   }
if(F22<=fmin3(F11,F12,F21))
   {
   if(F22<*F)
      {
      *X=X2;
      *Y=Y2;
      *F=F22;
      }
   if(n>0)
      bisect((Xm+Xx)/2.,Xx,(Ym+Yx)/2.,Yx,func,
  n-1,X,Y,F);
   }
}
```

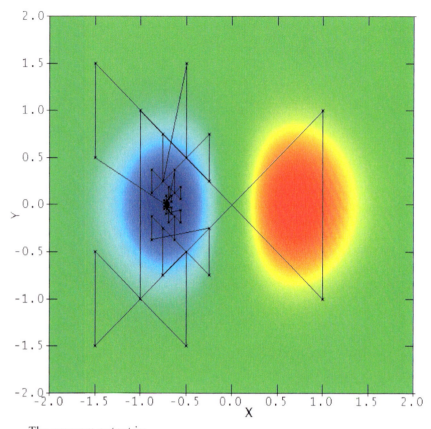

The program output is:

```
searching for solution
n=8, X=-1, Y=-1, F11=-1.21802
n=7, X=-0.5, Y=-0.5, F22=-2.72939
n=6, X=-0.75, Y=-0.25, F12=-3.61301
n=5, X=-0.625, Y=-0.125, F22=-3.74706
n=4, X=-0.6875, Y=-0.0625, F12=-3.84191
n=3, X=-0.71875, Y=-0.03125, F12=-3.85513
n=2, X=-0.703125, Y=-0.015625, F22=-3.85887
n=1, X=-0.710938, Y=-0.0078125, F12=-3.85959
n=0, X=-0.707031, Y=-0.00390625, F22=-3.85988
```

This technique is basically a bisection search in two dimensions. Consider the first example in Chapter 6, $z=(cosh(x-0.5)+(y-0.5)^2)/8$. The solution path for this problem (example2.c) is similar:

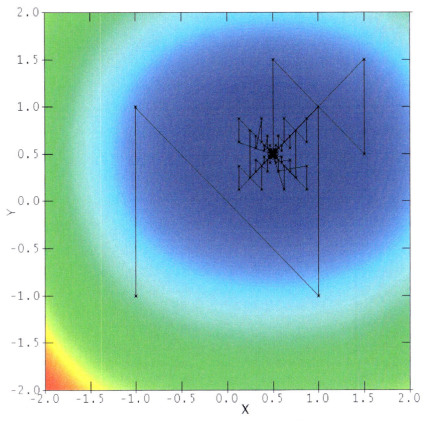

The last example of this technique we will consider is the first three-variable problem in Chapter 7: $f_1=15*x_1-cos(5/9*(x_2-1/5)*\pi*x_3)-1/2$, $f_2=25*x_1{}^2-81*(x_2-1/10)^2-sin(5*\pi*x_3/9)+1$, $f_3=exp(-5*x_1*(x_2-1/5))-100*\pi*x_3/9+10*\pi/3-1$. This requires expanding the technique to an arbitrary number of dimensions. The code (example3.c) must allocate and free memory every time it enters and leaves the function; otherwise, the values won't be preserved and the arrays will be orphaned in the heap.

Extension of this algorithm to an arbitrary number of dimensions requires a variable number of nested loops. While we could insert separate code for one loop, another for two, yet another for three, and so forth, but this would be tedious and unimaginative. What we need is a loop of loops! This is part of the test I give all prospective programmers. A professional should be able to implement it efficiently.

The most compact code performing an arbitrary loop of loops is the following:

```
do{
  for(i=0;i<n;i++)
    {
    if(k[i]<l[i]-1)
      {
      k[i]++;
      break;
      }
    k[i]=0;
    }
  }while(i<n);
```

Typical output is as follows:
```
loops 2 3 4
0 0 0 0
1 1 0 0
2 0 1 0
etc...
21 1 1 3
22 0 2 3
23 1 2 3
2*3*4=24
```

The loop of loops code, along with a batch file to compile and test it, can be found in the online archive in folder examples\bisection. We insert the loop of loops into the two-variable bisection code (example1.c and example2.c) to get the n-variable bisection code (example3.c). Then use this to solve the first three-variable problem from Chapter 7. Note the search pattern (1/4, 3/4), then split in half, and repeat:

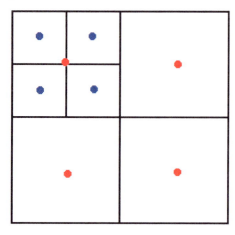

The 3D solution path for this problem is shown in the following figure:

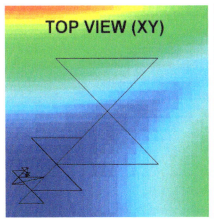
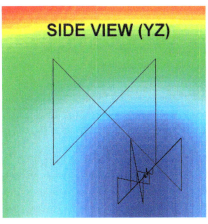
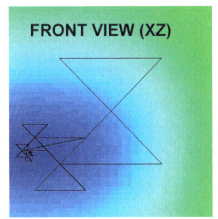

X=0.1
Y=0.2
Z=0.3

Chapter 10: Evolutionary Method

Evolutionary methods are based on the notion that accumulated small, random changes may eventually lead to something useful. The simplest possible implementation is to randomly hop around the domain, increasingly biasing the location with any fortuitous results. The following figure shows a one-dimensional example of this technique:

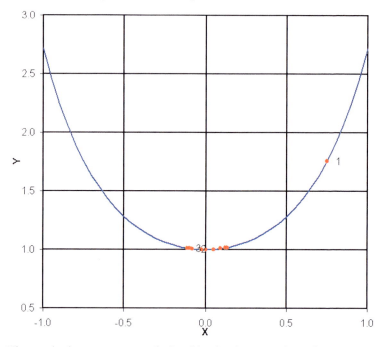

The method appears to work for this simple example. Of course, a rock could find this minimum.

The "bump" example from Chapter 3 is a little more challenging. It's shown in this next figure. The files can be found in the online archive in folder examples\evolutionary. This method isn't particularly efficient, but it's better than a completely random search.

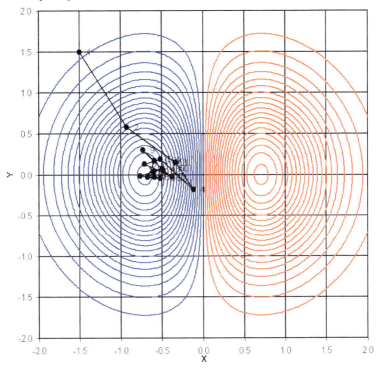

The third problem from Chapter 9 is an even greater challenge, as illustrated in this next figure:

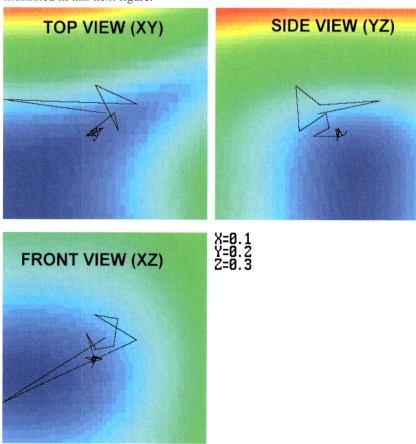

The method runs out of steam long before finding the blue core of the residual volume.

There are countless modifications, which might be applied to coax this method along. The simplest is to retard the influence of the previous solutions. This is illustrated in the next figure:

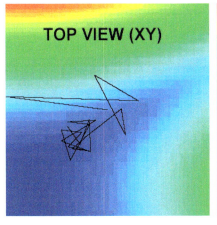
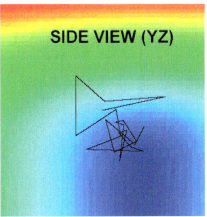
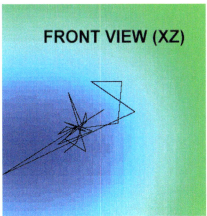

Retardation helps a little in this particular case, but we want methods that work for most cases, not ones that must be endlessly tweaked for every new application. We will consider one more tweak before moving on to the next example. In this modification, we save the *n* best locations, where *n* is the number of independent variables or the dimension of the solution space and bias the guesses with this retained information about the residual. We allocate space for *n+1* residuals (*Fj*) an locations (*Xj*). We start with *n* random guesses and sort these on the residual. Each update goes in locations *Fj[n]* and *Xj[n]* and the list get resorted. The following code (example3a.c) implements this technique:

```
void EvolutionarySearch(double*Xm,double*Xx,int
    n,double func(double*),int m,double*X,double*F)
```

```
{
int i,j;
double*Fj,*Xj;
Fj=calloc((n+1)*(n+2),sizeof(double));
Xj=Fj+n+1;
for(j=0;j<n;j++) /* n random initial guesses */
  {
  for(i=0;i<n;i++)
    Xj[n*j+i]=Xm[i]+rnd()*(Xx[i]-Xm[i]);
  Fj[j]=func(Xj+n*j);
  }
bubble_sort(Fj,Xj,n,n);
*F=Fj[0]; /* save best so far */
memcpy(X,Xj,n*sizeof(double));
for(;m>0;m--)
  {
  for(i=0;i<n;i++)
    {
    Xj[n*n+i]=n*(Xm[i]+rnd()*(Xx[i]-Xm[i])); /*
  random guess weighted with */
    for(j=0;j<n;j++) /* last n values with least Fjs
  */
      Xj[n*n+i]+=(n-j)*Xj[n*j+i];
    Xj[n*n+i]/=(n*(n+3))/2;
    }
  Fj[n]=func(Xj+n*n);
  bubble_sort(Fj,Xj,n,n+1);
  if(Fj[0]<*F)
    {
    *F=Fj[0]; /* save best so far */
    memcpy(X,Xj,n*sizeof(double));
    }
  }
free(Fj);
}
```

A bubble sort is good enough for our purposes here, as these arrays don't lend themselves to using the library function qsort(). The results are disappointing, as illustrated in the next figure. This modification utilizes more previous information and was allowed to run for five times as many iterations, yet failed to approach the solution. We have moved from one to two to three dimensions and already the Evolutionary Method seems to be petering out. This observation will be reinforced by subsequent examples comparing the gradient and evolutionary options of the Excel® Solver Add-In. Even after many, many more trial solutions, the gradient method stomps the evolutionary. Intelligence consistently beats brute force.

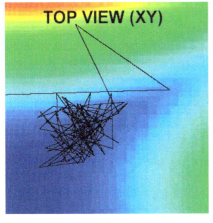
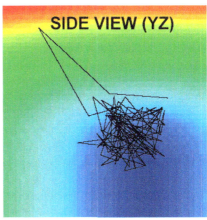

```
X=0.1
Y=0.2
Z=0.3
```

The Evolutionary Method quickly gets stuck in a region away from the solution where a steepest descent along the gradient would quickly lead to the optimal point. Of course, there are countless ways of restarting the process and trying over-and-over again, but you could apply those same "tricks" to one of the more intelligent methods (like I already did to Broyden's) and be ahead.

Chapter 11: Nonlinear Regression

In Chapter 5 we introduced nonlinear least squares, which is the basic strategy for nonlinear regression. In this chapter we will consider more complex problems. Linear regression is a non-iterative process in which matrices are built and solved, yielding a single result. Not that such are completely without complication, but iterative methods are much more complex and fraught with difficulties. We rarely approach a regression with no anticipation of the final form and so it is with the following examples.

Model Parameter Tuning

The first case we will consider is a mass transfer phenomenon that was killing fish in a similar way that the "bends" (i.e., decompression sickness) impacts divers. Some investigators thought this impossible, as fish move freely through the deep, but the biological evidence was overwhelming and persistent. The unusual situation impacting the fish was due to supersaturation of the water with nitrogen, immediately downstream of a dam.

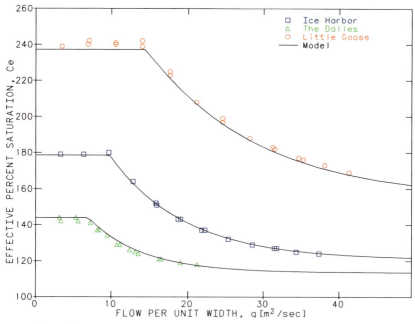

After collecting field data from four different sites (Ice Harbor, The Dalles, Little Goose, and Jennings-Randolph), a modeling effort was undertaken to better understand the mechanisms involved.[12] Supersaturation is a mass transfer process by which nitrogen from air bubbles enters the water and later tissues of

---

[12] Benton, D. J., "Modeling Nitrogen Supersaturation at Jennings-Randolph," U. S. Army Corps of Engineers Report. 1998.

the fish. Mass transfer processes are necessarily time-dependent and lead to the following solution:

$$C_d = C_s - (C_s - C_u)e^{-Kt} \qquad (11.1)$$

$$C_e = C_s \left(1 + \frac{d_e \gamma}{P_a}\right) \qquad (11.2)$$

$$d_e = h_2 + (h_1 - h_2)e^{\left(1 - \frac{\beta h_b}{L_s}\right)} \quad \text{for} \quad \frac{\beta h_b}{L_s} > 1$$

$$d_e = h_1 \quad \text{for} \quad \frac{\beta h_b}{L_s} \leq 1 \qquad (11.3)$$

The heights, depths, and lengths are the physical dimensions of the dam and river The rate of change of the concentration is given by:

$$\frac{dC}{dt} = K_1 a_b (C_e - C) + K_2 a_s (C_s - C) \qquad (11.4)$$

Here, K_1 and K_2 are mass transfer coefficients. The bubble mass transfer parameters are shown in this next figure:

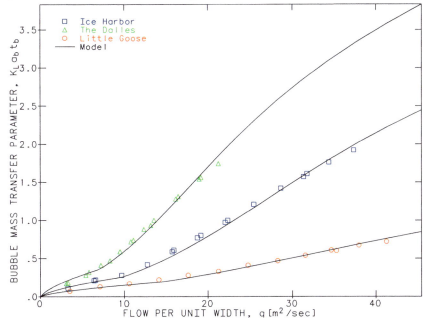

The solution to this differential equation is:

$$C_d = C_e - (C_e - C_u)\Omega$$
$$\Omega = e^{-(K_1 a_b t_b + K_2 a_s t_s)} + \Phi \left[1 - e^{-(K_1 a_b t_b + K_2 a_s t_s)}\right] \quad (11.5)$$
$$\Phi = \frac{(K_1 + K_2)a_s t_s}{K_1 a_b t_b + K_2 a_s t_s} \left(\frac{C_e - C_s}{C_e - C_u}\right)$$

The concentrations, C_d, C_e, and C_u, are unknown and will be adjusted to best fit the field data. As before, we have some idea what these must be, but the exact values are uncertain. This gives us bounds within which to search for an optimum solution, that is, one that best agrees with the available data. We will use Broyden's Method to find these three coefficients. All of the associated files can be found in the folder examples\nitrogen. The normalized optimum solution is shown in this next figure:

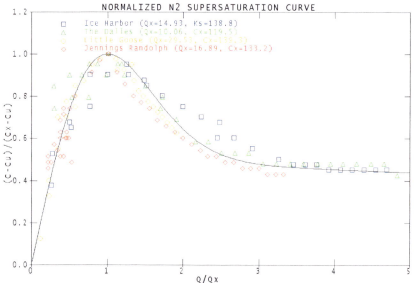

The residual function is simple and solved as in the previous examples using Broyden's Method:

```
double fun(double a,double b,double c,double x)
  {
  double C11,C12,C21,C22,D;
  C11=exp(-a)*erf(b);
  C12=erf(c);
  C21=exp(-a)*(a*erf(b)*sqrt(M_PI)-2*b*exp(-b*b));
  C22=-2*c*exp(-c*c);
  D=C11*C22-C21*C12;
  if(fabs(D)<FLT_MIN)
    Abort(__LINE__,"can't make match point");
  A=C22/D;
```

```
B=-C21/D;
return(A*exp(-a*x)*erf(b*x)+B*erf(c*x));
}
```

The particular (non-normalized) optimum solution is shown in this figure:

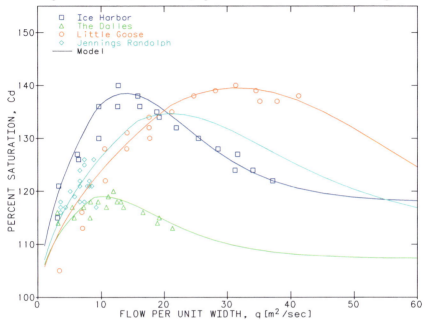

In case you're wondering... changes were made and the fish are much happier. My report convinced the right people that this was indeed the problem. The preceding equations were used to optimize the reconstruction efforts.

Combined Cycle Power Plant Performance Data

We will now consider a greater challenge: a pair of nonlinear regression problems with 10 unknowns and 8562 residuals. We seek two approximations (power and heat input) based on one year of data from a typical modern combined cycle power plant. The data, code, and associated files can be found in the online archive in folder examples\power plant data.

In this case we have very definite expectations regarding the form of the regressions. There is a great body of work on power plant performance. Two foundation references are ASME PTC-46 (Overall Plant Performance) and PTC-22 (Gas Turbines). Equipment manufacturers and plant designers all over the world follow these guidelines. Original equipment performance is most often provided as laid out in these two documents. Power and heat input are expected to have the following form:

$$\frac{P_{REF}}{P} = \prod_{i=1}^{14} \alpha_i \qquad (11.6)$$

$$\frac{Q_{REF}}{Q} = \prod_{i=1}^{14} \beta_i \qquad (11.7)$$

The power, P, and heat input, Q, are normalized by some reference value, typically the guarantee at the base reference conditions. The most common base reference conditions are 1 atmosphere (corrected for local elevation), 59°F (15°C), and 60% relative humidity. The multiplicative corrections, α_1 through α_{14} and β_1 through β_{14}, account for deviations from the base reference conditions. Each index corresponds to a specific parameter, as listed in the following table.

In this case we will only use the ones in bold: temperature, pressure, humidity, fuel composition, speed, equivalent hours of operation (EOH), and inlet guide vane position (indicative of part-load operation). The last two are not allowed in a contractual acceptance test, when demonstration of full-load operation in "new and clean" condition is the objective. Two can be quickly simplified: pressure and speed, as capacity (power) and consumption (heat input) are directly proportional to these. All of the corrections must be unity (i.e., equal to one) at the base reference conditions.

1	**inlet air temperature**
2	**barometric pressure**
3	**inlet humidity**
4	**fuel composition**
5	injection fluid flow
6	injection fluid enthalpy
7	injection fluid composition
8	exhaust pressure loss
9	**shaft speed**
10	turbine extraction
11	fuel temperature
12	inlet pressure loss
13	**hours of operation**
14	**inlet guide vane position**

The impact of temperature on gas turbine (and thus combined power plant) performance varies with design, but generally follows the same trend. These (and every other heat engine) perform better at low temperatures than high. As these corrections are the inverse, this means that α_1 is greater than 1 at low temperatures and less than 1 at high temperatures, as illustrated below:

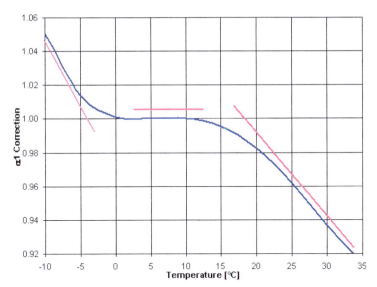

There is often (but not always) a flat spot in the middle of this curve. It is very important that the asymptotic behavior at both ends of the curve exhibit a negative slope. This means, if we are to approximate the shape of this curve as a polynomial in temperature, it must have an odd order of at least three. We rarely exceed order three, as the curve may flip around—something to avoid.

Relative humidity is a familiar term, but most people don't appreciate the sizeable impact water vapor content has on air. We all know that hot and humid is more uncomfortable than hot and dry. At 70°F (21°C) and 100% relative humidity, there is more energy in the water vapor than the dry air. At 120°F (49°C) and only 40% relative humidity, this is also true. The water vapor bearing capacity of air changes vastly over the annual range of ambient temperatures in most locations.

The impact of relative humidity on gas turbine (and thus combined cycle) performance is highly nonlinear and strongly dependent on temperature; however, the impact of *absolute* humidity is linear. Relative humidity is the fraction of water vapor compared to the maximum (i.e., bearing capacity); whereas, absolute humidity is simply the mass fraction, irrespective of the bearing capacity. This is why it's always preferable to base such regressions on absolute, rather than relative, humidity.

The impact of fuel composition (the chemistry of natural gas varies with the source) is most often adequately approximated as being linear in heating value and carbon to hydrogen ratio. The impact of ageing (equivalent hours of operation) is most pronounced early in the life of an engine and flattens-out over time. This particular engine has been running for several years, so we will assume it to be on the flat (i.e., linear) region.

Heat rate is the ratio of the heat input to the power and is inversely proportional to the thermal efficiency. Heat rate is a common measure of performance for power plants. Heat rate has considerably greater scatter than either power or heat input, which is why these two are calculated or measured before taking the ratio and why we will base our second regression on heat input rather than heat rate.

The corrections are as follows:

$$\alpha_1 = C_0 + C_1 T + C_2 T^2 + C_3 T^3 \tag{11.8}$$

$$\alpha_3 = C_0 + C_1 humidity \tag{11.9}$$

$$\alpha_4 = C_0 + C_1 LHV + C_2 CH_{RATIO} \tag{11.10}$$

$$\alpha_{13} = C_0 + C_1 EOH \tag{11.11}$$

$$\alpha_{14} = C_0 + C_1 IGV + C_2 IGV^2 \tag{11.12}$$

The β corrections for heat input have the same form. Each of these equations can be simplified by using the difference between actual and base reference quantities, which makes C_0 equal to 1, for instance, in the case of temperature:

$$\Delta_T = \frac{T - T_{BASE}}{T_{BASE}} \tag{11.13}$$

We also know from experience that inverting the 13th and 14th corrections results in less scatter, so that the final complete form of the regression becomes:

$$P = C_0 \left(\frac{baro}{baro_{BASE}} \right) \left(\frac{speed}{speed_{BASE}} \right) \times$$
$$\frac{(1 + C_4 \Delta_{LHV} + C_5 \Delta_{CH_{RATIO}})(1 + C_6 \Delta_T + C_7 \Delta_T^2 + C_8 \Delta_T^3)(1 + C_9 \Delta_{HUM})}{(1 + C_1 \Delta_{EOH})(1 + C_2 \Delta_{IGV} + C_3 \Delta_{IGV}^2)} \tag{11.14}$$

This gives us 10 parameters to adjust in order to minimize the residual (i.e., sum of the squares of the discrepancies) with respect to the reported data. The regression for heat input has the same form. This problem is nonlinear in that the coefficients multiply each other. We can separate the two groups on the bottom of Equation 11.14, thus:

- $$\frac{P(1 + C_1 \Delta_{EOH})(1 + C_2 \Delta_{IGV} + C_3 \Delta_{IGV}^2) baro_{BASE} speed_{BASE}}{C_0 (1 + C_4 \Delta_{LHV} + C_5 \Delta_{CH_{RATIO}})(1 + C_6 \Delta_T + C_7 \Delta_T^2 + C_8 \Delta_T^3)(1 + C_9 \Delta_{HUM})} = baro \times speed \times \tag{11.15}$$

This doesn't eliminate the nonlinearity. In fact, expanding Equation 11.15 results in 30 terms. Solving the problem using linear regression would require finding 30 coefficients and then requiring 20 combinations thereof to have certain behavior such that that the whole would reduce back down to Equation 11.15. As this would also be a nonlinear problem, we solve the problem as is.

The Excel® Solver Add-In was first used to determine the 10 coefficients. The Solver takes several minutes to converge to a solution with a residual of $10^{5.005}$ for power and $10^{6.541}$ for heat input. Broyden's derivative-free method (see file plant_data.c) is much faster (0.3 second) and achieves the same residual. The Solver algorithm is a Generalized Gradient Reduction (GRG) method, using finite differences to calculate the various partial derivatives. Of course, any native executable will run a lot faster than Excel®.

After running for about 1 second (the native, not Excel® version, which takes several minutes), the Evolutionary Method achieved a residual of $10^{6.309}$. After running for over 9 seconds the residual was finally reduced to $10^{5.775}$. Results of the power regression are shown in the following figure and of the heat input regression are shown in the one after that. Either solver can be selected in Excel®. The multi-dimensional bisection method achieved a residual of $10^{6.309}$ after running for about 9 seconds. All three methods are included in the source code.

The program output is:
```
examples\power plant data>plant_data
reading data: plant_data.csv
  8563 lines read
```

```
  8562 data points found
Power Regression
  Uncorrelated Data
    log10(residual)=9.017
  Excel GRG Solver
    log10(residual)=5.005
  Bisection Search
    9216 calls 8.507 seconds
    log10(residual)=5.555
  Broyden's Method
    100 calls 0.333 seconds
    log10(residual)=5.006
  Evolutionary Method
    1009 calls 0.942 seconds
    log10(residual)=6.309
  Evolutionary Method
    10009 calls 9.429 seconds
    log10(residual)=5.775
Heat Input Regression
  Uncorrelated Data
    log10(residual)=5.775
  Excel GRG Solver
    log10(residual)=6.541
  Bisection Search
    9216 calls 8.623 seconds
    log10(residual)=7.222
  Broyden's Method
    79 calls 0.225 seconds
    log10(residual)=6.542
  Evolutionary Method
    1009 calls 0.939 seconds
    log10(residual)=7.816
  Evolutionary Method
    10009 calls 9.304 seconds
    log10(residual)=7.453
```

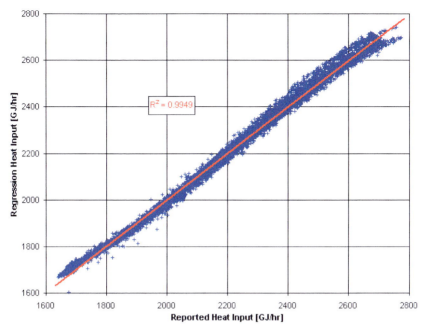

An important principle is illustrated by these examples: simplistic techniques based on wishful thinking or dependant on brute force are rarely preferable to a thoughtful and reasoned approach. One application where this wisdom has been embraced is computer chess. The most successful programs don't try to consider every possible move; rather, they use strategy. You will find that it take quite a while for the Evolutionary Method to run in Excel®. Try it and see!

Pressure-Temperature-Density Data

The next example we will consider is the pressure, temperature, and density behavior of real fluids. Every pure substance that is chemically stable over a sufficient range of pressure and temperatures exhibits both a critical and triple point. The critical point is where the liquid and vapor are physically indistinguishable. The triple point is where the solid, liquid, and vapor coexist in equilibrium. The critical point of water occurs at 22 MPa (3200 psia) and 374°C (705°F). The triple point occurs at 611 Pa (0.89 psia) and 0°C (32°F).

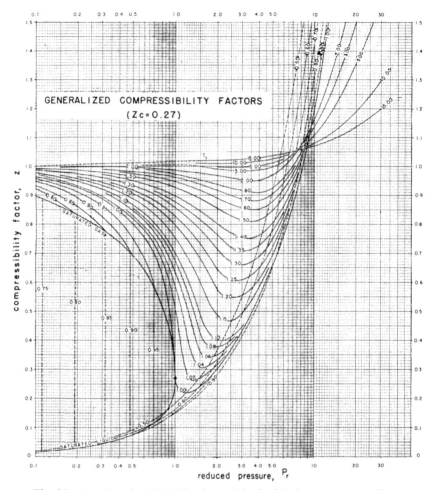

The ideal gas law is: $PV=RT$, where R is the ideal gas constant. For non-ideal states we define the compressibility factor, $Z=PV/RT$. Nelson and Obert introduced the generalized compressibility chart in 1954[13]. The preceding figure shows compressibility, Z, vs. reduced pressure, $Pr=P/Pcritical$, for various values of reduced temperature, $Tr=T/Tcritical$. I have digitized this figure and put the data into an Excel® spreadsheet in folder examples\compressibility. Fitting a single function to this entire family of curves might seem like an insurmountable task, but that's precisely what we're going to do.

---

[13] Nelson, L. C. and Obert, E. F., "Generalized pvT Properties of Gases," Transactions of the ASME, p. 1057, October, 1954.

The first equation of state to account for real fluid behavior was introduced by van der Waals in 1873: $P = RT/(V-b) - a/V^2$. V is the specific volume, equal to 1/density. The two constants, a and b, are selected to match the "bulge" in the preceding figure at $Pr=1$, $Z=0.27$. In order to form the "bulge" the following must be true:

$$\frac{\partial P}{\partial V} = \frac{\partial^2 P}{\partial V^2} = 0 \bigg|_{P=Pc, V=Vc} \quad (11.16)$$

While the van der Waals equation has somewhat the correct shape, it does not agree well with data, especially for liquids. We seek an improvement upon this first attempt at capturing real fluid behavior. The simplest improved form is:

$$P = \frac{RT}{V-b} - \frac{a}{\left(V^2 + cV + d^2\right)} \quad (11.17)$$

We will allow a, b, c, and d to be temperature-dependent. Examining the data reveals that a, c, and d should all get smaller with increasing temperature, while b should get smaller. Therefore, we propose the following relationships:

$$a, c, d \propto Tr^n, n > 0$$
$$b \propto Tr^n, n < 0 \quad (11.18)$$

This gives us 4 constants plus 4 exponents to adjust so as to minimize the residual. Experience also reveals that trying to match pressures at the same specific volume (1/density) yields spurious and problematic results. Instead, we want to match specific volume at the same pressure. This means that all of the calculations are implicit, as the X-axis of the figure is pressure, not specific volume. It takes some coding and it runs slowly, but Excel® can handle it and arrives at the following results. The blue specks are the digitized points and the red curves are the regression.

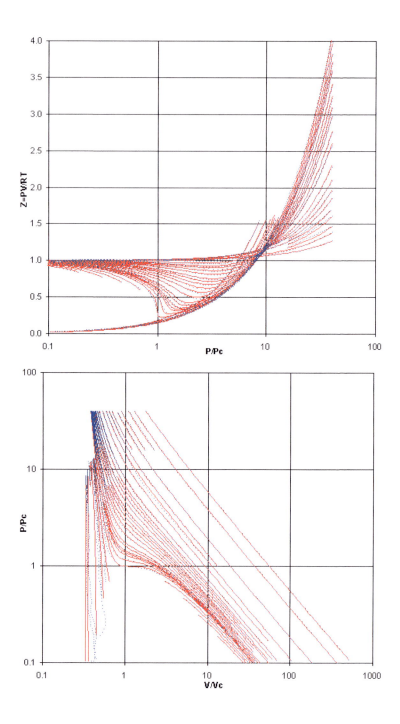

The second figures shows the same data on a different axis, this time reduced pressure vs. reduced volume. On the second figure, the critical point is at location (1,1) and also appears as a "hump" of sorts. Considering the X-axis in both figures is logarithmic, as is the Y-axis on the second figure, the agreement for such a simple equation is remarkable.

We apply the same techniques as for the power plant data (see source code in file compressibility.c). The program outputs:

```
examples\compressibility>compressibility
reading data: compressibility.csv
  3420 lines read
  3419 data points found
Uncorrelated Data
  residual=830114.310
Excel GRG Solver
  residual=1.019
Bisection Search
  2304 calls 7.143 seconds
  residual=7.643
Broyden's Method
  100 calls 0.386 seconds
  residual=1.060
Evolutionary Method
  1007 calls 3.056 seconds
  residual=4.223
Evolutionary Method
  10007 calls 30.391 seconds
  residual=1.304
```

Again, Broyden's Method is much faster than the Excel® GRG Method and yields equivalent results. The Bisection Method is slow and yields somewhat disappointing results. The Evolutionary Method is even slower and yields disappointing results.

Viscosity

Like thermodynamic properties, transport properties are essential to solving many scientific and engineering problems. The viscosity of liquids and vapors varies considerably over the range of operating temperatures. Perhaps the most studied fluid is water. As with preceding example, a high-order linear regression is unlikely to have the character we desire or the correct asymptotic behavior; thus, we turn again to nonlinear regression and an informed choice of form.

The following graph shows the dynamic viscosity of water in kg/s/m $\times 10^{-6}$, which is milli-centi-poise. Other than being convenient numbers ranging up to 1000, I have no idea why anyone would use such a combination of metric units. In any event, divide by 413.3788731 to get the far more useful units of lbm/ft/hr. You can find all of the related files in folder examples\viscosity.

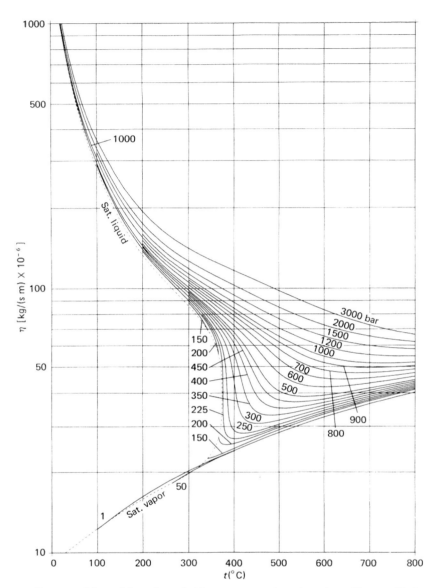

From working with various fluids, we expect that viscosity will more likely be related to density than either temperature or pressure, so we transform the X-axis into density [gm/cm²] and re-plot to obtain the following, much more tractable regression problem:

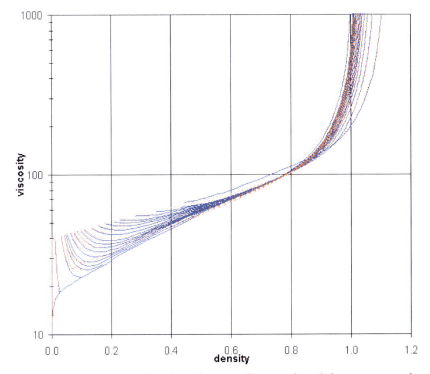

The blue curves are data from the preceding graph and the orange specks are the regression. Again, this is surprisingly good. The basic form is:

$$\mu = \frac{\alpha}{\delta - \rho}$$
$$\alpha = f(P, T, \rho) \quad (11.19)$$
$$\delta = f(P)$$

After trying several formulas, we arrive at the following expressions for α and δ:

$$\alpha = C_1 + C_2 P + C_3 T + C_4 \rho + C_5 P^2 \ldots$$
$$+ C_6 PT + C_7 P\rho + C_8 T^2 + C_9 T\rho + C_{10} \rho^2 \quad (11.20)$$
$$\delta = C_{11} + C_{12} P + C_{13} P^2$$

We adjust the 13 coefficients to minimize the residual. In this case, we want to minimize the relative discrepancies and so we calculate the residuals:

$$r_i = \frac{fit - data}{data} \quad (11.21)$$

The same behavior would result if taking the difference in the logs. We don't use the logs because Excel® gets upset when taking the log of a negative value and such may arise during the solution process. It takes a few minutes, but Excel® settles in on the solution displayed in the previous graph. The program (viscosity.c) is very similar to the previous (compressibility.c). The output is:

```
examples\viscosity>viscosity
reading data: viscosity.csv
  2289 lines read
  2287 data points found
Uncorrelated Data
  residual=2287.000
Excel GRG Solver
  residual=2.702
Bisection Search
  114688 calls 16.609 seconds
  residual=575.417
Broyden's Method
  100 calls 0.080 seconds
  residual=265.087
Broyden's second chance
  100 calls 0.083 seconds
  residual=3.281
Evolutionary Method
  1012 calls 0.154 seconds
  residual=8.806
Evolutionary Method
  10012 calls 1.460 seconds
  residual=4.704
```

Broyden's Method doesn't work well at all for this case (residual 265 vs. 2.7 for Excel's GTG solver) in spite of being fed a reasonable facsimile of the answer because the solution domain is radically split up where $\mu=a/(\delta-\rho)$ blows up. This method will find a solution if such regions are eliminated, as with the second chance above.

Chapter 12: Hybrid Regression

Sometimes part of the regression can be done linearly and part cannot. In this case, we repeatedly perform the linear least squares part and adjust the rest to minimize some residual. The problem we will consider is based on Case 4 in the SIAM paper. The objective is to develop a simple approximation that will reasonably approximate the transient behavior of a reservoir, including the time lag from the upstream and downstream dams. The upstream and downstream flows come from actual operations at the dam. The calculated flows come from a reservoir routing model, which is too large and takes too long to run to be embedded in the decision management software; thus, a simplified solution is required.

The simplified solution approximates the calculated flow from a linear combination of 4 flows: the current upstream and downstream value plus two previous flows, one from upstream and one from downstream. The time lag for each is sought so as to best approximate the actual flow. The time lag is presumed to be at least 1 and no more than 24 hours. The time lags for the upstream and downstream are presumed to be different.

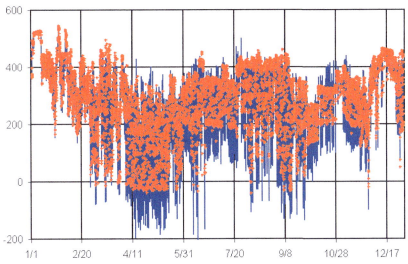

Integer lags are assumed in the Excel® spreadsheet (examples\ reservoir\ reservoir.xls) to facilitate solution, although this isn't strictly necessary. Excel's GRG solver doesn't work with this problem because it can't calculate the gradient of the residual by finite difference with respect to the time lags. Excel's Evolutionary solver does work, producing an answer of 2.33 (rounded to 2) and 6 hours for the upstream and downstream flows, respectively. The final residual is 15,407,749. I have provided a button to run through 24x24 cases sequentially, saving the best one, which turns out to be 3 and 7, with a residual of 12,682,871. The Evolutionary solver actually takes longer than the brute force double loop, as it tries more combinations.

The end result is shown in the preceding figure. The blue lines are the flows from the routing model and the orange specks are the approximation. We will implement this in C (reservoir.c) without restricting the time lags to be integers. This would be possible in Excel® but impractical. The program output is:

```
examples\reservoir>reservoir
reading data: reservoir.csv
  8784 lines read
  8784 data points found
  Uncorrelated Data
  log(residual)=8.943
begin brute force (loop) solution
  1 1 18237780
  2 1 14481130
  3 1 12708036
  3 6 12695179
  3 7 12682871
  576 calls 0.218 seconds
  log(residual)=7.103
Excel GRG Solver
  log(residual)=7.103
Bisection Search
  12 calls 0.006 seconds
  lu=3.9, ld=21.1, log(residual)=7.173
Broyden's Method
  100 calls 0.088 seconds
  lu=4.9, ld=8.9, log(residual)=7.107
Evolutionary Method
  1001 calls 0.435 seconds
  lu=3.1, ld=7.4, log(residual)=7.099
Evolutionary Method
  10001 calls 4.392 seconds
  lu=3.2, ld=19.1, log(residual)=7.122
```

All of the methods work fairly well and run a *whole lot faster* than Excel®.

Chapter 13. Data Analysis

The first application of numerical calculus to be considered is data analysis. In this first chapter the focus will be limited to evenly spaced data. The two most common measurements meeting this criterion are time (e.g., 0, 5, 10, 15, 20 seconds) and length (e.g., 0.1, 0.2, 0.3, 0.4 cm). The quantity measured is often temperature (degrees), volts, amps, watts, or grams. The analysis methods discussed in this chapter apply equally well to all of these measurements.

Several things may be extracted from this data that implicitly involve calculus, including: 1) the slope or rate of change, 2) the accumulation or integration, 3) the initial state (backward in time), and 4) the final state (forward in time). An example of the first category would be the rate of temperature change from which might be deduced thermal conductivity or insulating properties. An example of the second category would be the rate of deposition or filling from which we might deduce concentration or flow.

The third and fourth categories are often overlooked but are of significant practical interest. Sometimes it is not practical or even possible to measure the initial state of a system. Starting up an experiment may in itself be a disruption that interferes with initial measurements. It is often not practical to conduct very lengthy experiments. Both of these situations provide motivation for extrapolation, either backward or forward in time—something we can easily do with numerical calculus. This figure illustrates all four categories:

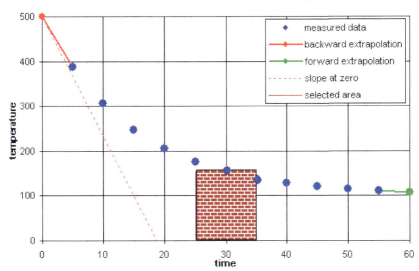

In this case, it's simple to come up with the first order relationships. For the temperature at t=0 sec and t=60 sec:

$$T_0 \approx 2T_5 - T_{10} \qquad (13.1)$$

$$T_{60} \approx 2T_{55} - T_{50} \tag{13.2}$$

The slope at t=0 is approximately the same as the slope at t=5, or:

$$\left.\frac{dT}{dt}\right|_{t=0} \approx \frac{T_{10} - T_5}{\Delta t} \tag{13.3}$$

where $\Delta t = 5$. The selected area is approximately:

$$A_{25-35} \approx T_{30}\, 2\Delta t \tag{13.4}$$

The second order approximations are a little more complicated:

$$T_0 \approx 3T_5 - 3T_{10} + T_{15} \tag{13.5}$$

$$T_{60} \approx T_{45} - 3T_{50} + 3T_{55} \tag{13.6}$$

$$\left.\frac{dT}{dt}\right|_{t=0} \approx \frac{-5T_5 + 8T_{10} - 3T_{15}}{2\Delta t} \tag{13.7}$$

$$A_{25-35} \approx (T_{25} + T_{35})\Delta t \tag{13.8}$$

Algebraic expressions for third order approximations are quite tedious to arrive at without a symbolic processor such as Maple®. While third order integration isn't a problem, extrapolation and differentiation are. These can have considerable error and be much less accurate than second and first order. Still, sometimes we would like to know these results. This problem naturally leads to the topic of smoothing.

It is possible to calculate the derivative using 4 points, which would be third order, but also require one degree of smoothing, effectively reducing this estimation to second degree, while achieving greater accuracy than second degree and taking an additional data point into consideration. It is also possible to consider 5 points plus 2 degrees of smoothing and so forth. Extrapolation can also be combined with smoothing and even more points so that many combinations are possible.

While such approximations are often tailored for each specific application, it is possible to implement this process in general. The usefulness of such an approach is readily apparent when applied to repetitive measurements or some sort of production process. Another application for this technique is process control—calculations that are carried out continuously.

The conventional approach for determining these coefficients for combined smoothing, integrating, differentiating, and extrapolating would be to set up a matrix called a Vandermonde and then invert it. The general form of such a matrix is as follows:

$$A = \begin{bmatrix} 1 & X_1 & X_1^2 & X_1^3 \\ 1 & X_2 & X_2^2 & X_2^3 \\ 1 & X_3 & X_3^2 & X_3^3 \\ 1 & X_n & X_n^2 & X_n^3 \end{bmatrix} \qquad (13.9)$$

This is how one typically fits a polygon (i.e., y=a+b*x+c*x²+d*x³) to a series of points and then integrates or differentiates the polygon. This sounds simple enough, but it quickly runs out of steam with large orders (e.g., x^{50}) due to accumulated round-off errors and limited precision floating-point calculations. Inverting a matrix is always subject to such errors.

Thankfully, there is a completely different approach to solving this problem that doesn't involve inverting a matrix and isn't subject to accumulated round-off error: orthogonal polynomials. Details of this approach are described and the source code listed in Appendix C. That program will be used here to obtain coefficients. For example:

```
transformation coefficients for equally-spaced data
p=points, i=point, d=differentiate (-1 for
   integrate),s=smoothing

p=2,i=0,d=0,s=0
2,-1           Equation 13.1

p=2,i=0,d=1,s=0
-1,1           Equation 13.3

p=3,i=0,d=0,s=0
3,-3,1         Equation 13.5

p=3,i=0,d=1,s=0
-2.5,4,-1.5    Equation 13.7
```

This program accommodates real numbers for parameter i so that you can estimate the value, slope, or integral at locations in between the points, for example midway between points 1 and 2 would be 1.5, as in:

```
transformation coefficients for equally-spaced data
p=points, i=point, d=differentiate (-1 for
   integrate),s=smoothing

p=3,i=1.5,d=0,s=0
0.375,0.75,-0.125 the value at 13.5

p=3,i=1.5,d=1,s=0
-1,1,0         the derivative at 13.5
```

```
p=3,i=1.5,d=-1,s=0
0.666667,0.416667,-0.0833333
```

This next sequence of coefficients reveals the impact of smoothing along with increasing order in estimating the value at zero (for which we don't have data):

```
transformation coefficients for equally-spaced data
p=points, i=point, d=differentiate (-1 for
    integrate),s=smoothing

p=2,i=0,d=0,s=0
2,-1

p=3,i=0,d=0,s=1
1.33333,0.333333,-0.666667

p=4,i=0,d=0,s=2
1,0.5,0,-0.5

p=5,i=0,d=0,s=3
0.8,0.5,0.2,-0.1,-0.4

p=6,i=0,d=0,s=4
0.666667,0.466667,0.266667,0.0666667,-0.133333,-
    0.333333

p=7,i=0,d=0,s=5
0.571429,0.428571,0.285714,0.142857,0,-0.142857,-
    0.285714

p=8,i=0,d=0,s=6
0.5,0.392857,0.285714,0.178571,0.0714286,-0.0357143,-
    0.142857,-0.25
```

Notice that each set of coefficients in this set sum to unity. This is always the case with extrapolation, interpolation, and integration. For differentiation the coefficients will always sum to zero.

Further Differentiation

If these coefficients with smoothing aren't adequate for your differentiation needs, I suggest you fit a curve through the data and analytically differentiate it. For more details, see my book entitled, *Curve-Fitting*:

http://www.amazon.com/dp/B01LDUK032

Chapter 14. Newton-Cotes Integration

The earliest method for numerical integration more sophisticated than the Trapezoidal Rule[14] is attributed to Isaac Newton[15] and Roger Coates[16]. This method uses equally-spaced points (i.e., $x_{i+1}-x_i$=constant) and can be represented by the following formula:

$$\int_a^b y\,dx \approx (b-a)\sum_{i=1}^{n} C_i y(x_i)$$
$$x_i = \frac{a(n-i)+b(i-1)}{(n-1)} \tag{14.1}$$

The method works fairly well for a small number of points but falls apart for large values of n. The reason this and some other methods *fall apart* is very important and why this method is presented here. The reasoning behind Newton-Cotes integration can be summarized as follows: If a polygon of order n-1 went through every one of the points (y_i) and that polygon was analytically integrated, what would be the result? The general polygon of order n-1 that goes through each y_i at x_i is the *Lagrange interpolator* and is given by the following formula:

$$p_n(x) = \sum_{i=1}^{n} \frac{y_i \prod_{\substack{j=1 \\ j \neq i}}^{n}(x-x_j)}{\prod_{\substack{j=1 \\ j \neq i}}^{n}(x_i-x_j)} \tag{14.2}$$

This formula can be expanded and integrated analytically in order to obtain the coefficients. The first few sets of coefficients are as follows:

$$n = 2 : \frac{1}{2}, \frac{1}{2} \tag{14.3}$$

$$n = 3 : \frac{1}{6}, \frac{2}{3}, \frac{1}{6} \tag{14.4}$$

$$n = 4 : \frac{1}{8}, \frac{3}{8}, \frac{3}{8}, \frac{1}{8} \tag{14.5}$$

$$n = 5 : \frac{7}{90}, \frac{32}{90}, \frac{12}{90}, \frac{32}{90}, \frac{7}{90} \tag{14.6}$$

---

[14] The Trapezoidal Rule is so elementary that it won't be covered here.
[15] Sir Isaac Newton 1642-1746: English mathematician, astronomer, theologian, author, and physicist.
[16] Roger Cotes 1682-1716: English mathematician, Fellow of the Royal Society.

Newton-Cotes for n=4 is often called Simpson's Rule[17] and n=5 is often called Boole's Rule[18]. The coefficients can be determined without solving a matrix. A program to do so is listed in Appendix D. The coefficients for orders up to 60 are listed there.

All of the coefficients are positive for orders less than or equal to 8. Beginning at order 9 one or more of the coefficients is negative. As these always sum to unity, this means that one or more of the coefficients must be larger than one for orders above 8. This leads to divergence.

With any such method, we would hope that increasing order would consistently increase accuracy, but this is not the case with this method due to this instability. In order to test the Newton-Cotes method, a simple integral with an analytical solution is chosen:

$$\int_0^\pi \sin(x)^2 = \frac{\pi}{2} \qquad (14.6)$$

The results for orders 3 through 60 are as follows:
```
order 3: 2.094395
order 4: 1.767146
order 5: 1.535890
order 6: 1.551975
order 7: 1.572666
order 8: 1.571918
order 9: 1.570724
order 10: 1.570751
order 11: 1.570798
order 12: 1.570798
order 16: 1.570796
order 20: 1.570796
order 30: 1.570796
order 40: 1.570796
order 50: 1.570797
order 60: 1.565138
exact: 1.570796
```

For this example, we achieve seven digits of accuracy with orders 16 through 40 and then lose accuracy for higher orders. The test program (NCtest.c) may also be found in the examples\ NewtonCotes folder.

Extended Rules

Simpson's Rule (a.k.a. Newton-Cotes for n=4) is often extended in some repeating pattern. We will consider such methods here, as they are of historical interest; however, they are of little value in practice. Far more efficient methods

---

[17] Thomas Simpson (1710-1761): English mathematician.
[18] George Boole (1815-1864): English mathematician, educator, philosopher, and logician.

are available that provide increased accuracy with fewer function evaluations, thus less computing time. We will consider several variants. The first of these might be called the 1-4-2 Rule, as the coefficients are 1 on each end plus alternating 4 and 2 in between. For 5 points this becomes: 1,4,2,4,1 and for 7 points becomes: 1,4,2,4,2,4,1. The second variant might be called the 1-3-2 Rule. For 5 points this is: 1,3,2,3,1 and for 7 points: 1,3,2,3,2,3,1. Divide by the sum in each case to normalize. The code to perform this task is rather simple:

```
#define _CRT_SECURE_NO_DEPRECATE
#include <stdio.h>
#include <stdlib.h>
#define _USE_MATH_DEFINES
#include <float.h>
#include <math.h>
double f(double x)
  {
  return(pow(sin(x),2));
  }
double Simpson(double f(double x),double a,double b)
  {
  return((b-
   a)*(f(a)+3.*f((a+2.*b)/3.)+3.*f((2.*a+b)/3.)+f(b))/8
   .);
  }
double Simpson132(int n,double f(double x),double
   a,double b)
  {
  int i;
  double r,s;
  if(n<5||n%2==0)
    return(-999.);
  s=f(a)+f(b);
  r=2.;
  for(i=2;i<=n-1;i+=2)
    {
    s+=3.*f((a*(n-i)+b*(i-1))/(n-1))+2.*f((a*(n-i-
    1)+b*i)/(n-1));
    r+=3.+2.;
    }
  return((b-a)*s/r);
  }
double Simpson142(int n,double f(double x),double
   a,double b)
  {
  int i;
  double r,s;
  if(n<5||n%2==0)
    return(-999.);
  s=f(a)+f(b);
  r=2.;
```

```
  for(i=2;i<=n-1;i+=2)
    {
    s+=4.*f((a*(n-i)+b*(i-1))/(n-1))+2.*f((a*(n-i-
  1)+b*i)/(n-1));
    r+=4.+2.;
    }
  return((b-a)*s/r);
  }
int main(int argc,char**argv,char**envp)
  {
  int n;
  printf("Testing Simpson's Rules\n");
  printf("4 pt: %lf\n",Simpson(f,0.,M_PI));
  for(n=5;n<=61;n+=n>21?4:2)
    printf("%i pt: %lf or %lf\n",n,
      Simpson132(n,f,0.,M_PI),
      Simpson142(n,f,0.,M_PI));
  printf("exact: %lf\n",M_PI/2.);
  return(0);
  }
```

The results for the same integral $(\sin(x)^2)$ are listed below:

```
Testing Simpson's Rules
4 pt: 1.767146
5 pt: 1.308997 or 1.346397
7 pt: 1.385997 or 1.413717
9 pt: 1.427997 or 1.449966
11 pt: 1.454441 or 1.472622
13 pt: 1.472622 or 1.488123
15 pt: 1.485888 or 1.499396
17 pt: 1.495997 or 1.507964
19 pt: 1.503954 or 1.514696
21 pt: 1.510381 or 1.520125
23 pt: 1.515681 or 1.524596
27 pt: 1.523907 or 1.531526
31 pt: 1.529996 or 1.536649
35 pt: 1.534686 or 1.540589
39 pt: 1.538409 or 1.543714
43 pt: 1.541436 or 1.546253
47 pt: 1.543945 or 1.548356
51 pt: 1.546059 or 1.550128
55 pt: 1.547865 or 1.551640
59 pt: 1.549425 or 1.552946
exact: 1.570796
```

The performance is quite poor, which is why nobody uses either one of these methods. There are some diehards that won't let go of Simpson's method and so they came up with yet another variant using subdivision. This algorithm is even easier to implement with a reentrant call: split the interval in half and apply the 1,3,3,1 Rule to each half, making 8 points. You can split each of those intervals in half, making it 16 points, and so forth. Here's the code:

```
#define _CRT_SECURE_NO_DEPRECATE
#include <stdio.h>
#include <stdlib.h>
#define _USE_MATH_DEFINES
#include <float.h>
#include <math.h>
double f(double x)
  {
  return(pow(sin(x),2));
  }
double SplitSimpson(int n,double f(double x),double
    a,double b)
  {
  if(n)
    return(SplitSimpson(n-
    1,f,a,(a+b)/2.)+SplitSimpson(n-1,f,(a+b)/2.,b));
  return((b-
    a)*(f(a)+3.*f((a+2.*b)/3.)+3.*f((2.*a+b)/3.)+f(b))/8
    .);
  }
int main(int argc,char**argv,char**envp)
  {
  int n;
  printf("Testing Split Simpson's Rule\n");
  for(n=0;n<=5;n++)
    printf("%i pt:
    %lf\n",4<<n,SplitSimpson(n,f,0.,M_PI));
  printf("exact: %lf\n",M_PI/2.);
  return(0);
  }
```

This works much better than the extended Simpson Rules, as you can see:

```
Testing Split Simpson's Rule
4 pt: 1.767146
8 pt: 1.570796
16 pt: 1.570796
32 pt: 1.570796
64 pt: 1.570796
128 pt: 1.570796
exact: 1.570796
```

Chapter 15. Gauss Quadrature

Gauss[19] Quadrature[20] is the premier method for numerical integration. Unlike many other methods (e.g., Newton-Cotes), it just keeps getting better as the number of points increases. Not only does the convergence increase with more points, the rate of convergence also increases—in practice, if not in theory. Gauss Quadrature (GQ) of order n will precisely integrate any polynomial up to order n+1.

The biggest difference between GQ and other methods (with the exception of Chebyshev and Lobatto Quadrature, which we will discuss in the next chapter) is the x-values (i.e., the location of the points) are also selected so as to improve the result. In the context of numerical integration or quadrature, the coefficients are often called *weights* and the points are often called *abscissas*. The Newton-Cotes abscissas were evenly-spaced and the weights were determined by requiring the exact integral of any polynomial up to some order. The same is true for GQ, but the addition of optimal point selection add twice as many conditions per order. This addition eliminates the problem of negative weights and divergence.

Orthogonal Polynomials

The details of orthogonal polynomials are beyond the scope of this work, although these were introduced in Chapter 13. Those polynomials were orthogonal when summed up at evenly-spaced points. The polynomials we're interested in here are orthogonal when integrated over the interval from -1 to +1, or:

$$\int_{-1}^{+1} P_i(x) P_j(x) = \frac{2}{2n+1}, i = j$$
$$\int_{-1}^{+1} P_i(x) P_j(x) = 0, i \neq j$$
(15.1)

The polynomials that satisfy this condition are called Legendre[21]. These can be generated through several relationships, including the following:

$$P_n(x) = \frac{1}{2^n} \sum_{k=0}^{n} B(n,k) B\left(\frac{n+k-1}{n}, n\right) x^k \qquad (15.2)$$

where *B(i,j)* is the binomial coefficient. More often these are generated by the recursion relationship:

$$(n+1) P_{n+1}(x) = (2n+1) x P_n(x) - n P_{n-1}(x) \qquad (15.3)$$

---

[19] Carl Friedrich Gauss (1777-1855): German mathematician.
[20] The term *quadrature* comes from making squares as in drawing boxes to implement the Trapezoidal Rule for integration.
[21] Adrien-Marie Legendre (1752-1833): French mathematician.

where $P_0(x)=1$ and $P_1(x)=x$. The first 6 are shown in the following figure:

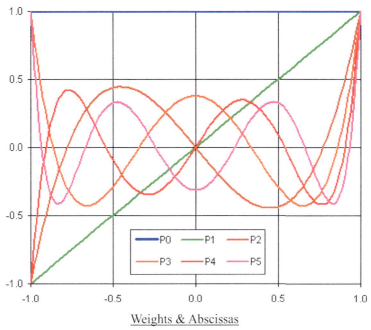

Weights & Abscissas

We could go through the lengthy process of setting up the simultaneous equations to find the n weights and n abscissas such that any polynomial of order up to order 2n would be precisely integrated over the interval from -1 to +1. These simultaneous equations would be nonlinear due to the powers of x_i. Solution of such equations would be tedious and have limited accuracy. There is no point in doing this because the solution may be found from the properties orthogonality of the Legendre polynomials.

There are n roots (locations where the polynomial passes through y=0) for the n+1 order Legendre polynomial. These are the abscissas. If you plug these values into the n order Legendre polynomial, you get the weights. The proof is quite involved and not the primary concern here. The GQ abscissas and weights for orders 3 through 6 are listed here and for orders 7 through 12 are listed in Appendix E. Abscissas and weights for orders up to 4096 may be found in the examples\GaussQuadrature folder in the on-line archive.

```
double A3[]={0.000000000000000,0.774596669241483};
double W3[]={0.888888888888888,0.555555555555556};
double A4[]={0.339981043585,0.861136311594};
double W4[]={0.652145154863,0.347854845137};
double A5[]={0.000000000000,0.538469310106,
    0.906179845939};
double W5[]={0.568888888889,0.478628670499,
    0.236926885056};
```

```
double A6[]={0.238619186083,0.661209386466,
   0.932469514203};
double W6[]={0.467913934573,0.360761573048,
   0.171324492379};
```

The code to implement GQ is very simple. There's a slight difference for odd orders, as the central abscissa is zero and the weight is only used once. The rest are trivial to implement. Here's the function:

```
double GQuad(double*A,double*W,int n,double f(double
   x),double a,double b)
  {
  int i;
  double ax,dx,q;
  ax=(a+b)/2.;
  dx=(b-a)/2.;
  if(n%2)
    {
    q=W[0]*f(ax);
    for(i=1;i<(n+1)/2;i++)
      q+=W[i]*(f(ax-dx*A[i])+f(ax+dx*A[i]));
    }
  else
    {
    for(q=i=0;i<n/2;i++)
      q+=W[i]*(f(ax-dx*A[i])+f(ax+dx*A[i]));
    }
  return(q*dx);
  }
```

You will find a program (GQtest.c) in the examples\GaussQuadrature folder, most of which is data statements. It calculates this same integral $(\sin(x)^2)$ for orders up to 4096. Here are the results:

```
Testing Gauss Quadrature
order 2:   1.19283364802896540
order 3:   1.60606730241801920
order 4:   1.56911897502940610
order 5:   1.57084471252998000
order 6:   1.57079538896380870
order 7:   1.57079633985864310
order 8:   1.57079632665773450
order 9:   1.57079632679478530
order 10:  1.57079632679580520
order 12:  1.57079632679293630
order 16:  1.57079632679489660
order 20:  1.57079632679605700
order 32:  1.57079632679489660
order 40:  1.57079632679387940
order 64:  1.57079632679489630
order 128: 1.57079632679489630
order 256: 1.57079632679489700
```

```
order 512: 1.57079632679489610
order 1024: 1.57079632679489610
order 2048: 1.57079632679489570
order 4096: 1.57079632679489660
exact: 1.57079632679489660
```

Order 16 achieves the exact solution to 18 significant figures! This is why I say GQ is the *premier* method. In this same folder you will find a little program to test GQ (tank.c) by calculating the volume of an ellipsoidal tank and hemispherical end caps. The implementation is simple. Here are the functions to be integrated as before:

```
double HorizontalCylinder(double h)
  {
  double w,x;
  if(h<0.)
    return(0.);
  x=2.*r-h;
  if(x<=0.)
    return(0.);
  w=2.*sqrt(h*x);
  return(w*l);
  }
double HorizontalElipsoidalCap(double h)
  {
  double a,w,x;
  if(h<0.)
    return(0.);
  x=2.*r-h;
  if(x<=0.)
    return(0.);
  w=2.*sqrt(h*x);
  a=b*w/r/2.;
  return(M_PI*a*w/4.);
  }
double GaussQuadrature(double F(double),double a,double
   b)
  {
  int i;
  double cx,dx,q;
  cx=(a+b)/2;
  dx=(b-a)/2;
  for(q=i=0;i<sizeof(A)/sizeof(A[0]);i++)
    q+=W[i]*(F(cx-dx*A[i])+F(cx+dx*A[i]));
  return(q*dx);
  }
```

The function calls to perform the integrals also simple:
```
GaussQuadrature(HorizontalCylinder,0.,h)
GaussQuadrature(HorizontalElipsoidalCap,0.,h)
```

20-point GQ is adequate, as illustrated by the following results:

```
horizontal cylinder volume
r l  h   analytic  numeric
1 2 0.5    1.228    1.228
1 2 1.3    4.323    4.323
1 4 0.5    2.457    2.457
1 4 1.3    8.647    8.647
1 6 0.5    3.685    3.685
1 6 1.3   12.970   12.970
2 2 0.5    1.813    1.813
2 2 2.1   13.366   13.366
2 2 3.7   24.276   24.277
2 4 0.5    3.626    3.627
2 4 2.1   26.732   26.733
2 4 3.7   48.553   48.554
2 6 0.5    5.440    5.440
2 6 2.1   40.098   40.099
2 6 3.7   72.829   72.831
3 2 0.5    2.251    2.251
3 2 2.9   27.075   27.075
3 2 5.3   52.860   52.861
3 4 0.5    4.502    4.502
3 4 2.9   54.149   54.150
3 4 5.3  105.720  105.723
3 6 0.5    6.752    6.752
3 6 2.9   81.224   81.225
3 6 5.3  158.580  158.584
horizontal elipsoidal cap
r  b   h   analytic  numeric
1 0.5 0.5    0.164    0.164
1 0.5 1.3    0.752    0.752
2 0.5 0.5    0.180    0.180
2 0.5 2.1    2.251    2.251
2 0.5 3.7    4.122    4.122
2 1.0 0.5    0.360    0.360
2 1.0 2.1    4.503    4.503
2 1.0 3.7    8.243    8.243
2 1.5 0.5    0.540    0.540
2 1.5 2.1    6.754    6.754
2 1.5 3.7   12.365   12.365
3 0.5 0.5    0.185    0.185
3 0.5 2.9    4.477    4.477
3 0.5 5.3    9.070    9.070
3 1.0 0.5    0.371    0.371
3 1.0 2.9    8.954    8.954
3 1.0 5.3   18.140   18.140
3 1.5 0.5    0.556    0.556
3 1.5 2.9   13.431   13.431
3 1.5 5.3   27.210   27.210
3 2.0 0.5    0.742    0.742
3 2.0 2.9   17.907   17.907
```

```
3 2.0 5.3  36.279  36.279
3 2.5 0.5   0.927   0.927
3 2.5 2.9  22.384  22.384
3 2.5 5.3  45.349  45.349
```

Open vs. Closed Interval

Notice that -1 and +1 are not included among the abscissas of any order GQ, although the end points do get closer as the order increases. Thus GQ is called an *open* interval method. The end points are included in *closed* methods. The Newton-Cotes (NC) Rules presented in Chapter 14 were all closed interval. There are open interval NC Rules, but these weren't covered because they have so little practical value. There are advantages to having open interval methods, as we will see in Chapter 18. Always use GQ when an open interval method is required.

Chapter 16. Chebyshev & Lobatto Quadrature

These two GQ-related methods are worth mentioning. Chebyshev Quadrature (CQ) is similar to GQ except all the weights are the same. Lobatto Quadrature (LQ) is the same as GQ except that -1 and +1 are the endpoints. CQ is an open method and LQ is closed. When an open method is needed, use GQ. When a closed method is needed and accuracy is desired, use LQ. I can't think of any reason to use CQ, although it has been used.[22]

Weights and abscissas for both Chebyshev and Lobatto Quadrature may be found in the *Handbook of Mathematical Functions* by Abramowitz and Stegun.[23] The abscissas and procedure for implementing Chebyshev Quadrature is quite simple:

```
/* Chebyshev abscissas */
double C2[]={0.5773502692};
double C3[]={0.7071067812};
double C4[]={0.1875924741,0.7946544723};
double C5[]={0.3745414096,0.8324974870};
double C6[]={0.2666354015,0.4225186538,0.8662468181};
double C7[]={0.3239118105,0.5296518105,0.8838617008};
double C9[]={0.1679061842,0.5287617831,0.6010186554,
   0.9115893077};
double Chebyshev(double*C,int n,double f(double
   x),double a,double b)
  {int i;
  double ax,dx,q;
  ax=(a+b)/2.;
  dx=(b-a)/2.;
  if(n%2)
    {
    q=f(ax);
    for(i=0;i<n/2;i++)
      q+=f(ax-dx*C[i])+f(ax+dx*C[i]);
    }
  else
    {
    for(q=i=0;i<n/2;i++)
      q+=f(ax-dx*C[i])+f(ax+dx*C[i]);
    }
```

---

[22] Merkel's method for integrating cooling tower demand uses 4-point Chebyshev Quadrature, perhaps for ease of graphical representation or ease of hand calculations in the days before computers. Merkel, F. Verdunstungskulung, V.D.I. Forschungsarbeiteh (Society of German Engineers Technical Journal), No. 275, Berlin, 1925.

[23] Abramowitz, M. and I. A. Stegun, *Handbook of Mathematical Functions* first published by the National Bureau of Standards as Technical Monograph No. 55. This very useful reference may be obtained free on-line as a PDF from several different web sites.

```
    return((b-a)*q/n);
    }
```

The weights, abscissas, and procedure for implementing Lobatto Quadrature is also quite simple:

```
/* Lobatto weights and abscissas */
double A3[]={0.5,1.};
double W3[]={4./3.,1./3.};
double A4[]={0.44721360,1.00000000};
double W4[]={0.83333333,0.16666667};
double A5[]={0.00000000,0.65465367,1.00000000};
double W5[]={0.71111111,0.54444444,0.10000000};
double A6[]={0.28523152,0.76505532,1.00000000};
double W6[]={0.55485838,0.37847496,0.06666667};
double A7[]={0.00000000,0.46884879,0.83022390,
   1.00000000};
double W7[]={0.48761904,0.43174538,0.27682604,
   0.04761904};
double A8[]={0.20929922,0.59170018,0.87174015,
   1.00000000};
double W8[]={0.41245880,0.34112270,0.21070422,
   0.03571428};
double A9[]={0.0000000000,0.3631174638,0.6771862795,
   0.8997579954,1.0000000000};
double W9[]={0.3715192744,0.3464285110,0.2745387126,
   0.1654953616,0.0277777778};
double A10[]={0.1652789577,0.4779249498,0.7387738651,
   0.9195339082,1.0000000000};
double W10[]={0.3275397612,0.2920426836,0.2248893420,
   0.1333059908,0.0222222222};
double Lobatto(double*A,double*W,int n,double f(double
   x),double a,double b)
  {int i;
  double ax,dx,q;
  ax=(a+b)/2.;
  dx=(b-a)/2.;
  if(n%2)
    {
    q=W[0]*f(ax);
    for(i=1;i<(n+1)/2;i++)
      q+=W[i]*(f(ax-dx*A[i])+f(ax+dx*A[i]));
    }
  else
    {
    for(q=i=0;i<n/2;i++)
      q+=W[i]*(f(ax-dx*A[i])+f(ax+dx*A[i]));
    }
  return(q*dx);
  }
```

A program to test each of these (CLtest.c) may be found in the examples\Lobatto folder of the on-line archive. Here are the results:

```
Testing Chebyshev Quadrature
order 2: 1.19283364797927430
order 3: 1.46010768479294240
order 4: 1.59614564019955290
order 5: 1.58292576263350940
order 6: 1.56933737248257190
order 7: 1.56999618095089670
order 9: 1.57083236167838860
Testing Lobatto Quadrature
order 3: 2.09439510239319530
order 4: 1.52507871486619420
order 5: 1.57286574390776800
order 6: 1.57073868336479850
order 7: 1.57079740532856140
order 8: 1.57079633358758320
order 9: 1.57079632717886280
order 10: 1.57079632665996140
exact: 1.57079632679489660
```

The Chebyshev is more often than not superior to Newton-Cotes with the same number of points and the Lobatto is almost as accurate as Gauss Quadrature.

Closed Interval

There are a few problems when the end points are critical and a closed method is needed. In these cases, Lobatto is preferred over Gauss. One such case is heat exchanger analysis. A temperature *pinch* often occurs at the inlet or exit of a heat exchanger. In such cases, the *tightness* of the *pinch* dominates the performance or dictates the size and expense.

Evaporative cooling towers are an example of this situation. The approach to the wet-bulb temperature (which is a type of pinch) drives the cost more than anything else. A close approach tower cost a lot more than a long approach one. It was previously mentioned that Merkel chose the 4-point Chebyshev method to integrate the governing differential equation and produce cooling tower demand curves. Chebyshev is not a closed method and does not include the end points or the pinch. Lobatto would have been a far better choice for this task, as I have demonstrated.[24]

---

[24] Benton, D. J., "Comparison of Methods for Numerical Integration in Computing Cooling Tower Demand Curves," Cooling Technology Institute, Performance Committee Report, Dana Point, California, 1989.

Chapter 17. Composite Rules

We have already considered some composite rules, including the extended and subdivided Simpson. The same type of subdivision with reentrant calls can be used to easily increase the number of points for most any other method. This will usually improve the accuracy, but doesn't necessarily improve the efficiency of the calculation.

Romberg's Method

At this point we mention Romberg's[25] method. The most significant feature of this method is that it provides an estimate of the error, which can be used to decide how many subdivisions are enough. It has very little practical value, as it takes a lot of points and there are so many more efficient methods available. Here is the code if you want to try it out:

```
double Romberg(double F(double),double X1,double X2,int
   n)
 {
 int i,i1,j,j1,jj;
 double dX,E,H,H2,S,X;
 static double R[15][15];
 dX=X2-X1;
 H=dX;
 R[0][0]=(F(X1)+F(X2))*H/2.;
 if(n<2)
   return(R[0][0]);
 jj=1;
 i1=0;
 for(i=1;i<n;i++)
   {
   S=0.;
   H2=H/2.;
   X=X1-H2;
   for(j=0;j<jj;j++)
     {
     X+=H;
     S+=F(X);
     }
   R[i][0]=(R[i1][0]+H*S)/2.;
   H=H2;
   E=1.;
   j1=0;
   for(j=1;j<=i;j++)
     {
     E*=4.;
     R[i][j]=(E*R[i][j1]-R[i1][j1])/(E-1.);
     j1=j;
     }
```

---

[25] Werner Romberg (1909-2003): German mathematician and physicist.

```
        jj+=jj;
        i1=i;
        }
    return(R[n-1][n-1]);
    }
```

In the examples\composite folder you will find a program (testall.c) that compares 75 methods for speed and accuracy, including: trapezoidal, Simpson, Newton-Cotes, Romberg, Chebyshev, Lobatto, and Gauss. The test function is the integral of 1/x from 1 to 2, which is ln(2). Here are the results:

testing various methods for numerical integration				
method	result	error	seconds	digits/sec
exact=ln(10)	2.30258509299404	N/A	N/A	N/A
3-point trapezoidal	3.29318181818181	0.990596725187773	0.000004	1,012
5-point trapezoidal	2.62922118204376	0.326636089049717	0.000007	72,890
10-point trapezoidal	2.37896825396825	0.076383160974208	0.000015	76,794
20-point trapezoidal	2.32071372799289	0.018128634998853	0.000027	65,311
100-point trapezoidal	2.30326634427572	0.000681251281679	0.000037	84,445
1000-point trapezoidal	2.30259178882424	0.000006695830197	0.000470	11,009
10,000-point trapezoidal	2.30258515983237	0.000000066838326	0.005320	1,349
3-point Simpson	2.74090909090909	0.438323997915045	0.000002	228,356
5-point Simpson	2.40790096999774	0.105315877003698	0.000002	391,002
10-point Simpson	2.32057573532623	0.017990642332192	0.000005	372,257
20-point Simpson	2.30399679153549	0.001411698541445	0.000010	285,026
100-point Simpson	2.30258741788446	0.000002324890419	0.000040	140,840
1000-point Simpson	2.30258509321443	0.000000000220390	0.000460	20,993
10,000-point Simpson	2.30258509299405	0.000000000000006	0.005310	2,675
4-point Newton-Cotes	2.56339285714285	0.260807764148811	0.000002	280,896
5-point Newton-Cotes	2.38570042860365	0.083115335609608	0.000003	411,872
6-point Newton-Cotes	2.35981628020375	0.057231187209712	0.000003	396,005
7-point Newton-Cotes	2.32471556804540	0.022130475051354	0.000003	474,436
8-point Newton-Cotes	2.31875667290220	0.016171579908159	0.000004	425,421
9-point Newton-Cotes	2.30946438340678	0.006879290412737	0.000005	459,522
10-point Newton-Cote	2.30777965561224	0.005194562618199	0.000005	442,612
11-point Newton-Cote	2.30492526495466	0.002340171960618	0.000005	491,074
12-point Newton-Cote	2.30438957505264	0.001804482058597	0.000006	445,843
16-point Newton-Cote	2.30283669104050	0.000251598046457	0.000009	404,920
20-point Newton-Cote	2.30262462735199	0.000039534357950	0.000010	440,303
30-point Newton-Cote	2.30258560470063	0.000000511706588	0.000014	461,338
40-point Newton-Cote	2.30258510168987	0.000000008695833	0.000020	403,034
50-point Newton-Cote	2.30258542275379	0.000000329759748	0.000023	283,579
60-point Newton-Cote	2.29959125650675	0.002993836487292	0.000032	78,868
3-point Romberg	2.74090909090909	0.438323997915045	0.000001	243,579
5-point Romberg	2.38570042860365	0.083115335609608	0.000003	398,368

Method	Value	Error	Time	Count
9-point Romberg	2.31362792006895	0.011042827074904	0.000005	391,384
17-point Romberg	2.30341497733484	0.000829884340796	0.000009	346,611
33-point Romberg	2.30261516949073	0.000030076496687	0.000017	271,306
65-point Romberg	2.30258555868970	0.000000465695660	0.000032	197,872
129-point Romberg	2.30258509581234	0.000000002818299	0.000053	160,313
257-point Romberg	2.30258509300029	0.000000000006244	0.000080	140,056
513-point Romberg	2.30258509299405	0.000000000000004	0.000150	96,330
1025-point Romberg	2.30258509299404	0.000000000000002	0.000470	31,384
2049-point Romberg	2.30258509299405	0.000000000000010	0.000940	14,904
4097-point Romberg	2.30258509299404	0.000000000000003	0.002030	7,179
8193-point Romberg	2.30258509299402	0.000000000000021	0.004210	3,247
16,385-point Romberg	2.30258509299402	0.000000000000020	0.008440	1,624
3-point Chebyshev	2.18520609828096	0.117378994713080	0.000002	569,876
4-point Chebyshev	2.25523458707396	0.047350505920079	0.000002	662,338
5-point Chebyshev	2.27069435812671	0.031890734867336	0.000003	570,478
6-point Chebyshev	2.28819920995952	0.014385883034520	0.000003	575,645
7-point Chebyshev	2.29231812702023	0.010266965973816	0.000004	534,425
9-point Chebyshev	2.29895639355763	0.003628699436416	0.000004	553,123
3-point Lobatto	2.74090909090909	0.438323997915045	0.000002	228,356
4-point Lobatto	2.39942749601619	0.096842403022152	0.000002	443,596
5-point Lobatto	2.32615591796764	0.023570824973597	0.000003	600,187
6-point Lobatto	2.30856660691480	0.005981513920759	0.000003	770,705
7-point Lobatto	2.30413500794603	0.001549914951991	0.000004	772,665
8-point Lobatto	2.30299146266670	0.000406369672661	0.000004	826,575
9-point Lobatto	2.30269242163076	0.000107328636723	0.000005	868,281
10-point Lobatto	2.30261357341938	0.000028480425339	0.000005	939,394
2-point Gauss	2.10638297872340	0.196202114270641	0.000001	654,249
3-point Gauss	2.24660974384731	0.055975349146734	0.000002	782,502
4-point Gauss	2.28696952387280	0.015615569121244	0.000002	745,157
5-point Gauss	2.29828311073711	0.004301982256929	0.000002	946,533
6-point Gauss	2.30140808410775	0.001177008886287	0.000003	915,381
7-point Gauss	2.30226434828873	0.000320744705316	0.000004	938,970
8-point Gauss	2.30249790203241	0.000087190961628	0.000004	964,138
9-point Gauss	2.30256142936713	0.000023663626913	0.000004	1,048,542
10-point Gauss	2.30257867788627	0.000006415107776	0.000006	941,194
12-point Gauss	2.30258462257900	0.000000470415039	0.000006	988,675
16-point Gauss	2.30258509048285	0.000000002511189	0.000008	1,021,264
20-point Gauss	2.30258509297903	0.000000000015010	0.000010	1,082,363
40-point Gauss	2.30258509299538	0.000000000001342	0.000020	593,619
96-point Gauss	2.30258509299404	0.000000000000003	0.000040	362,686
2*5-point Gauss	2.30232304578739	0.000262047206652	0.000005	671,554
10*5-point Gauss	2.30258508304717	0.000000009946871	0.000025	320,093
20*5-point Gauss	2.30258509296621	0.000000000027832	0.000053	197,915

| 100*5-point Gauss | 2.30258509299289 | 0.000000000001151 | 0.000160 | 74,619 |

As mentioned previously, Gauss Quadrature is by far the most efficient with Lobatto and Chebyshev following it. The subdivided methods (e.g., 2*5, 10*5, 25*5, and 100*5 point Gauss) offer no real improvement in performance, even for this simple example. No doubt, other problems would lead to somewhat different results, but I've investigated this and the same trends and generalizations are persistent.

Chapter 18. Green's Lemma

Green's Lemma[26] is usually covered in advanced calculus. This useful theorem transforms an area (2D) integral into a boundary integral (1D). It is often used in developing finite element solutions, which is where I have employed it with good success. The Lemma can be expressed by the following integral:

$$\iint \left(\frac{\partial f}{\partial x} - \frac{\partial g}{\partial y} \right) dx\,dy = \oint \left(f\,dx + g\,dy \right) \quad (18.1)$$

While it can quite complicated to integrate a function over an irregular area (e.g., finite elements), it may not be nearly as complicated to integrate around a boundary. The spanning or basis functions used in finite elements are often simple and can be analytically integrated over one or the other dimension. This step provides the transformation from $\partial f/\partial x \rightarrow f(x,y)$ and $\partial g/\partial y = g(x,y)$.

Integrating around a boundary (most often a polygon) is quite simple. Furthermore, this often doesn't require a high degree of quadrature. In the case of finite element basis functions, 4-point Gauss Quadrature is adequate. The code to implement this is simple:

```
double GreensLemma(double*Xp,double*Yp,int np,double
   U,double V)
{
static double
 A[]={0.339981043584856,0.861136311594053};
static double
 W[]={0.652145154862546,0.347854845137454};
double dX,dY,S,Xa,Xb,X1,X2,Ya,Yb,Y1,Y2;
int i,j;
X2=Xp[np-1];
Y2=Yp[np-1];
for(S=i=0;i<np;i++)
  {
  X1=X2;
  Y1=Y2;
  X2=Xp[i];
  Y2=Yp[i];
  dX=(X2-X1)/2.;
  dY=(Y2-Y1)/2.;
  Xb=(X2+X1)/2.;
  Yb=(Y2+Y1)/2.;
  for(j=0;j<2;j++)
    {
    Xa=pow(Xb-A[j]*dX,U);
    Ya=pow(Yb-A[j]*dY,V+1.);
```

---

[26] George Green (1729-1841): British mathematical physicist best known for work with electric fields and magnetism.

```
      S+=W[j]*dX*Xa*Ya;
      Xa=pow(Xb+A[j]*dX,U);
      Ya=pow(Yb+A[j]*dY,V+1.);
      S+=W[j]*dX*Xa*Ya;
      }
   }
   return(S/(V+1.));
}
```

In this case we are integrating the basis function X^U*Y^V. Below is a typical finite element grid that we will use as an example:

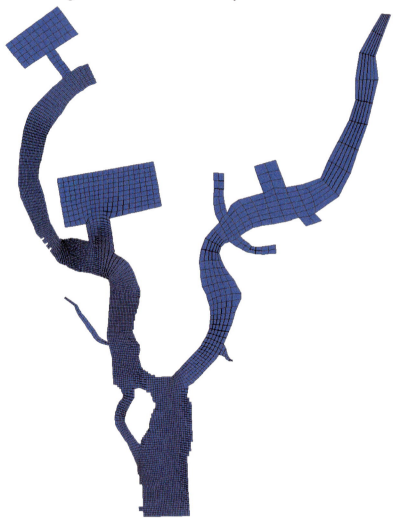

This FEM grid contains 5852 nodes and 5383 elements and may be found in the examples\GreensLemma folder in file grid.2dv. The program to perform the integrations is also in this same folder. Selected results from the calculations are listed below:

```
reading grid: grid.2dv
  5852 nodes
  5383 elements
 U    V    elem    integral
0.5  0.5     41  -7.07E+013
0.5  0.5   2318  -8.30E+013
0.5  0.5    951   2.36E+012
0.5  0.5   4968  -1.19E+014
0.5  0.5   3020  -8.20E+013
0.5  1.0   4958  -1.81E+017
0.5  1.0    712  -1.08E+017
0.5  1.0   2443  -1.30E+017
0.5  1.0     47  -1.19E+017
0.5  1.0   2932  -1.41E+017
0.5  1.5    322  -2.01E+020
0.5  1.5   1230  -2.15E+020
0.5  1.5   1749  -1.76E+020
0.5  1.5    678  -1.07E+020
0.5  1.5   4578  -2.50E+018
0.5  2.0    491  -3.69E+023
0.5  2.0   2995  -3.92E+023
0.5  2.0   1176  -4.08E+023
0.5  2.0   4827  -3.73E+023
0.5  2.0     53  -4.91E+023
1.0  0.5     93  -4.48E+016
1.0  0.5   3838   2.55E+017
1.0  0.5   3902   7.09E+016
1.0  0.5    153  -4.76E+016
1.0  0.5    292  -4.96E+016
```

The simplest application of Green's Lemma is computing the area of a polygon by integrating around the perimeter. As this is a first order process, the Trapezoidal Rule is adequate and produces exact results. The code is very simple:

```
double polygon_area(double*Xp,double*Yp,int np)
  {
  int i,j;
  double a;
  for(a=j=0,i=np-1;j<np;i=j++)
    a+=(Xp[j]-Xp[i])*(Yp[i]+Yp[j])/2.;
  return(fabs(a));
  }
```

Chapter 19. 2D and 3D Integrals

Sometimes it's necessary to integrate over a 2D or 3D domain. This is easily accomplished with most any of the quadrature formulas. I always use Gauss Quadrature because it's fast and accurate. General 2D integration can be expressed by the following integral:

$$\int_a^b \int_{g(x)}^{h(x)} f(x,y) dx\, dy \qquad (19.1)$$

The lower and upper limits of the inner integral are often functions of x and of the outer integral are often constants, but the same logic can be extended to handle functional outer limits of integration. The code to accomplish a 2D integral is quite simple:

```
double GQ2D(double f(double,double),double
   g(double),double h(double),double a,double b)
  {
  int i,j,k;
  double cx,cy,dx,dy,q,w,x,y1,y2;
  cx=(a+b)/2;
  dx=(b-a)/2;
  q=0;
  for(i=0;i<sizeof(A)/sizeof(A[0]);i++)
    {
    for(k=-1;k<=1;k+=2)
      {
      x=cx+k*dx*A[i];
      y1=g(x);
      y2=h(x);
      cy=(y1+y2)/2;
      dy=(y2-y1)/2;
      w=dy*W[i];
      for(j=0;j<sizeof(A)/sizeof(A[0]);j++)
        q+=w*W[j]*(f(x,cy-dy*A[j])+f(x,cy+dy*A[j]));
      }
    }
  return(q*dx);
  }
int n,m;
double f(double x,double y){return(pow(x,m)*pow(y,n));}
double g(double x){return(0.);}
double h(double x){return(M_PI);}
```

As a test case we will consider the following double integral:

$$\int_0^\pi \int_0^\pi x^n y^m dx\, dy = \frac{\pi^{2+n+m}}{(n+1)(m+1)} \qquad (19.2)$$

The code (test2D.c) to implement and test this integration, comparing it to the analytical solution, may be found in the examples\2Dplus folder of the on-line archive. The results are as follows:

```
testing 2D integration
n m analytic numerical
0 0    9.869604    9.869604
0 1   15.503138   15.503138
0 2   32.469697   32.469697
0 3   76.504921   76.504921
0 4  192.277839  192.277839
1 0   15.503138   15.503138
1 1   24.352273   24.352273
1 2   51.003281   51.003281
1 3  120.173649  120.173649
1 4  302.029323  302.029323
2 0   32.469697   32.469697
2 1   51.003281   51.003281
2 2  106.821022  106.821022
2 3  251.691102  251.691102
2 4  632.568734  632.568734
3 0   76.504921   76.504921
3 1  120.173649  120.173649
3 2  251.691102  251.691102
3 3  593.033189  593.033189
3 4 1490.454967 1490.454967
4 0  192.277839  192.277839
4 1  302.029323  302.029323
4 2  632.568734  632.568734
4 3 1490.454967 1490.454967
4 4 3745.921899 3745.921899
```

Recall that n-point Gauss Quadrature will precisely integrate any polynomial up to order n+1. As the highest order in this test is 9 (4+4+1) and 8-point GQ is used, all of the numerical results match the analytical to machine precision. In general, 3D integration can be represented by the following formula:

$$\int_a^b \int_{h1(x,y)}^{h2(x,y)} \int_{g1(x)}^{g2(x)} f(x,y,z) \, dx \, dy \, dz \quad (19.3)$$

The corresponding 3D quadrature is as follows:

```
double GQ3D(double f(double,double,double),double
    g1(double),double g2(double),double
    h1(double,double),double h2(double,double),double
    a,double b)
    {
    int i,j,k,l,m;
    double cx,cy,cz,dx,dy,dz,q,wz,wy,x,y,y1,y2,z1,z2;
    cx=(a+b)/2;
    dx=(b-a)/2;
```

```
   q=0;
   for(i=0;i<sizeof(A)/sizeof(A[0]);i++)
     {
     for(l=-1;l<=1;l+=2)
       {
       x=cx+l*dx*A[i];
       y1=g1(x);
       y2=g2(x);
       cy=(y1+y2)/2;
       dy=(y2-y1)/2;
       wy=dy*W[i];
       for(j=0;j<sizeof(A)/sizeof(A[0]);j++)
         {
         for(m=-1;m<=1;m+=2)
           {
           y=cy+m*dy*A[j];
           z1=h1(x,y);
           z2=h2(x,y);
           cz=(z1+z2)/2;
           dz=(z2-z1)/2;
           wz=dz*W[j];
           for(k=0;k<sizeof(A)/sizeof(A[0]);k++)
             q+=wy*wz*W[k]*(f(x,y,cz-
   dz*A[k])+f(x,y,cz+dz*A[k]));
           }
         }
       }
     }
   return(q*dx);
   }
int l,m,n;
double f(double x,double y,double
    z){return(pow(x,l)*pow(y,m)*pow(z,l));}
double g1(double x){return(0.);}
double g2(double x){return(M_PI);}
double h1(double x,double y){return(0.);}
double h2(double x,double y){return(M_PI);}
```

The 3D results are similar to the 2D:

```
testing 3D integration
l m n   analytical   numerical
0 0 0     31.00628    31.00628
0 0 1     48.70455    31.00628
0 0 2    102.00656    31.00628
0 0 3    240.34730    31.00628
0 1 0     48.70455    48.70455
0 1 1     76.50492    48.70455
0 1 2    160.23153    48.70455
0 1 3    377.53665    48.70455
0 2 0    102.00656   102.00656
```

```
0 2 1    160.23153    102.00656
0 2 2    335.58814    102.00656
0 2 3    790.71092    102.00656
0 3 0    240.34730    240.34730
0 3 1    377.53665    240.34730
0 3 2    790.71092    240.34730
0 3 3   1863.06871    240.34730
```

Chapter 20. Improper Integrals

Improper integrals have one or more limits that are infinite. There are many such integrals throughout applied mathematics. In general, we will consider the following form:

$$\int_0^\infty f(x)dx \qquad (20.1)$$

Other forms (such as from -infinity to +infinity) can easily be recast into this form algebraically, so that there's little point covering those specifically. In Chapter 16 we mentioned the distinction between open and closed quadrature forms. Here is where this important distinction comes into play. We will solve these integrals numerically using *open* quadrature, specifically Gaussian. Consider the following two transformations:

$$\int_0^\infty f(x)dx = \int_0^1 f\left(\frac{1}{y}-1\right)\frac{dy}{y^2} \qquad (20.2)$$

$$\int_0^\infty f(x)dx = \int_0^1 f(-\ln(y))\frac{dy}{y} \qquad (20.3)$$

We must use open quadrature because we don't want to evaluate the function at either 0 or 1 exactly. With open quadrature, we don't have to because the abscissas never get there. The implementation is simple:

```
double GQ0Ia(double*A,double*W,int n,double f(double))
  {/* using Equation (20.2)  */
  int i;
  double q,x;
  for(q=i=0;i<n/2;i++)
    {
    x=(1.-A[i])/2.;
    q+=W[i]*f(1./x-1.)/x/x;
    x=(1.+A[i])/2.;
    q+=W[i]*f(1./x-1.)/x/x;
    }
  return(q/2.);
  }
double GQ0Ib(double*A,double*W,int n,double f(double))
  {/* using Equation (20.3)  */
  int i;
  double q,x;
  for(q=i=0;i<n/2;i++)
    {
    x=(1.-A[i])/2.;
    q+=W[i]*f(-log(x))/x;
    x=(1.+A[i])/2.;
    q+=W[i]*f(-log(x))/x;
    }
```

```
  return(q/2.);
  }
double f(double x)
  {
  return(1./(1.+x)/sqrt(x));
  }
```

The code (improper.c) is in the examples\improper folder in the on-line archive. It uses the same weights and abscissas as GQtest.c discussed in Chapter 15. The test problem is:

$$\int_0^\infty \frac{dx}{(1+x)\sqrt{x}} = \pi \qquad (20.4)$$

The results are as follows:

```
Testing Improper Integrals
order       Equation (20.2)         Equation (20.3)
    2  2.44948974278317790   1.79448086891000670
    4  2.75540447558789210   2.06234879066470530
    6  2.87395041793634260   2.17672878406026050
    8  2.93684206475868060   2.24276607150958720
   10  2.97580942795443630   2.28679370330868090
   12  3.00231931310612850   2.31875310599846610
   16  3.03607059068002320   2.36290361690231700
   20  3.05665569109650680   2.39264451109131880
   32  3.08801383542212050   2.44505559952257820
   40  3.09859670208230930   2.46602976782676150
   64  3.11459478032419000   2.50393541705813270
  128  3.12804109039140290   2.54837757177866250
  256  3.13480365100789230   2.58353263249498650
  512  3.13819483897739330   2.61242140200313600
 1024  3.13989291694179280   2.63686388042320230
 2048  3.14074257780349560   2.65800067436659050
 4096  3.14116756381579340   2.67658002342123250
exact  3.14159265358979310
```

Not surprisingly, it takes a lot of points (a high degree of quadrature) to achieve good agreement with the analytical solution. As high order Newton-Cotes Rules (open or closed) are unstable, this precludes using them. We will, however, consider a subdivided approach. The subdivided Simpson code discussed in Chapter 16 can easily be adapted to accommodate the Gaussian abscissas and weights. The function can readily be transformed to implement Equations 20.2 and 20.3 so that the subdivided method can be evaluated. A snippet of the full code (impsplit.c) is listed below:

```
double SplitGauss(double*A,double*W,int n,int m,double
   f(double x),double a,double b)
   {
   int i;
   double ax,dx,q;
```

```
    if(m)
      return(SplitGauss(A,W,n,m-
    1,f,a,(a+b)/2.)+SplitGauss(A,W,n,m-1,f,(a+b)/2.,b));
    ax=(a+b)/2.;
    dx=(b-a)/2.;
    for(q=i=0;i<n/2;i++)
      q+=W[i]*(f(ax-dx*A[i])+f(ax+dx*A[i]));
    return(q*dx);
    }
double g(double x)
    {
    return(1./(1.+x)/sqrt(x));
    }
double fa(double y)
    {
    return(g(1./y-1.)/y/y);
    }
double fb(double y)
    {
    return(g(-log(y))/y);
    }
int main(int argc,char**argv,char**envp)
    {
    int m;
    printf("Split 8-pt Gauss Quadrature\n");
    printf("point   Eqn(20.2)  Eqn(20.3)\n");
    for(m=0;m<=10;m++)
       printf("%5i %9.7lf
      %9.7lf\n",8<<m,SplitGauss(A8,W8,8,m,fa,0.,1.),
      SplitGauss(A8,W8,8,m,fb,0.,1.));
    printf("Split 16-pt Gauss Quadrature\n");
    printf("point   Eqn(20.2)  Eqn(20.3)\n");
    for(m=0;m<=10;m++)
       printf("%5i %9.7lf
      %9.7lf\n",16<<m,SplitGauss(A16,W16,16,m,fa,0.,1.),
      SplitGauss(A16,W16,16,m,fb,0.,1.));
    printf("Split 32-pt Gauss Quadrature\n");
    printf("point   Eqn(20.2)  Eqn(20.3)\n");
    for(m=0;m<=10;m++)
       printf("%5i %9.7lf
      %9.7lf\n",32<<m,SplitGauss(A32,W32,32,m,fa,0.,1.),
      SplitGauss(A32,W32,32,m,fb,0.,1.));
    printf("exact   %9.7lf\n",M_PI);
    return(0);
    }
```

The results are as follows:

```
Split 8-pt Gauss Quadrature
point   Eqn(20.2)  Eqn(20.3)
    8 2.9368421 2.2427661
```

```
   16 2.9967531 2.3119032
   32 3.0391547 2.3672321
   64 3.0691506 2.4122695
  128 3.0903658 2.4495589
  256 3.1053688 2.4809500
  512 3.1159782 2.5077959
 1024 3.1234804 2.5310930
 2048 3.1287853 2.5515812
 4096 3.1325365 2.5698141
 8192 3.1351890 2.5862102
Split 16-pt Gauss Quadrature
point  Eqn(20.2)  Eqn(20.3)
   16 3.0360706 2.3629036
   32 3.0669692 2.4087139
   64 3.0888230 2.4465886
  128 3.1042779 2.4784282
  256 3.1152068 2.5056220
  512 3.1229349 2.5291928
 1024 3.1283996 2.5498992
 2048 3.1322637 2.5683088
 4096 3.1349961 2.5848497
 8192 3.1369282 2.5998490
16384 3.1382944 2.6135586
Split 32-pt Gauss Quadrature
point  Eqn(20.2)  Eqn(20.3)
   32 3.0880138 2.4450556
   64 3.1037056 2.4771285
  128 3.1148021 2.5045031
  256 3.1226488 2.5282160
  512 3.1281972 2.5490356
 1024 3.1321207 2.5675365
 2048 3.1348949 2.5841524
 4096 3.1368566 2.5992140
 8192 3.1382438 2.6129760
16384 3.1392246 2.6256373
32768 3.1399182 2.6373547
exact 3.1415927
```

The same pattern holds in that it takes a lot of subdivisions to achieve the accuracy of a higher order method. The first transformation of variables (Equation 20.2) continues to work considerably better than the second (Equation 20.3). The former trend is persistent, but the latter is not always the case. Consider two more transformations:

$$\int_0^\infty f(x)dx = \int_0^1 f\left(\frac{y}{1+y}\right)\frac{dy}{(1+y)^2} \tag{20.5}$$

$$\int_0^\infty f(x)dx = \int_0^{\frac{\pi}{2}} f(\tan(y))\frac{dy}{\cos^2 y} \tag{20.6}$$

The code to implement these is also simple:
```
double GQ0Ic(double*A,double*W,int n,double f(double))
   {
   int i;
   double q,x,y,z;
   for(q=i=0;i<n/2;i++)
      {
      x=(1.-A[i])/2.;
      y=1.+x;
      z=x/y;
      q+=W[i]*f(z)/y/y;
      x=(1.+A[i])/2.;
      y=1.+x;
      z=x/y;
      q+=W[i]*f(z)/y/y;
      }
   return(q);
   }
double GQ0Id(double*A,double*W,int n,double f(double))
   {
   int i;
   double c,d,q,x;
   d=M_PI/4.;
   for(q=i=0;i<n/2;i++)
      {
      x=(1.-A[i])*d;
      c=cos(x);
      q+=W[i]*f(tan(x))/c/c;
      x=(1.+A[i])*d;
      c=cos(x);
      q+=W[i]*f(tan(x))/c/c;
      }
   return(q*d);
   }
```

The program (improper2.c) is compiled and run as before to yield the following results:

```
Testing Improper Integrals
order      Equation 20.5        Equation 20.6
    2 1.71597303267771030 2.22603042920561030
    4 2.06807309586066480 2.65084467337283500
    6 2.19191241793135210 2.80409336963305570
    8 2.25613427256502770 2.88407785764290740
   10 2.29559248245833110 2.93334314579584450
   12 2.32232529582326920 2.96676191112965390
   16 2.35625829270656870 3.00922043971767030
   20 2.37690984517079330 3.03507760555438870
   32 2.40832198945200520 3.07442596377131050
   40 2.41891357351424170 3.08769716442585110
   64 2.43491865828793900 3.10775384473867430
  128 2.44836698031676820 3.12460803595272200
  256 2.45512979549372410 3.13308386900697580
  512 2.45852101548014530 3.13733412052996390
 1024 2.46021909745901100 3.13946234912195220
 2048 2.46106875882368610 3.14052724152694120
 4096 2.46149374489835670 3.14105988255817480
exact 3.14159265358979310
```

Here we see that the transformation in Equation 20.6 is far superior in this case to the one in Equation 20.5.

Chapter 21 Applications

For the remainder of this book we will consider various applications of the principles already presented. Applied mathematics is integral to the sciences, engineering, and even medicine. I trust this assortment of applications will illustrate the power of numerical calculus.

Debye Function

The Debye[27] function or integral arises in the field of thermodynamics and solid-state physics. Peter Debye first encountered it in 1912 while estimating the phonon contribution to the specific heat capacity of a solid. It is given by:

$$D_n(x) = \frac{n}{x^n} \int_0^x \frac{t^n}{e^t - 1} dt \qquad (21.1)$$

It can also be expressed by the following infinite series:

$$D_n(x) = 1 - \frac{n}{2(n+1)}x + n\sum_{k=1}^{\infty} \frac{B_{2k}}{(2k+n)(2k)!} x^{2k} \qquad (21.2)$$

where B_{2k} are the Bernoulli numbers. While it is possible to generate the Bernoulli numbers and sum up this infinite series until the factorial takes over and renders additional terms unnecessary, it is rather tedious. It is much simpler to solve this integral using quadrature and we already have sufficient orders of Gauss Quadrature to do so with precision. The GQ function below has been modified to pass an additional parameter, n (see Eqn. 9.1):

```
double GQuad(double*A,double*W,int m,double
   f(int,double),int n,double a,double b)
{
int i;
double ax,dx,q;
ax=(a+b)/2.;
dx=(b-a)/2.;
for(q=i=0;i<m/2;i++)
   q+=W[i]*(f(n,ax-dx*A[i])+f(n,ax+dx*A[i]));
return(q*dx);
}
double f(int n,double x)
{
return(pow(x,n)/(exp(x)-1.));
}
double Debye(int n,double x)
{
if(x<=0.)
   return(1.);
```

---

[27] Peter Debye (1884-1966): Dutch-American physicist and physical chemist; Nobel laureate in Chemistry.

```
    return(n*GQuad(A4096,W4096,4096,f,n,0.,x)/pow(x,n));
}
```

The results are more than adequate. This same table below is listed in Chapter 27 of Abramowitz & Stegun. You will also find a snapshot of that page in the applications\Debye folder of the on-line archive.

```
Debye Function
  x      n=1      n=2      n=3      n=4
  0.0  1.000000 1.000000 1.000000 1.000000
  0.1  0.975278 0.967083 0.963000 0.960555
  0.2  0.951111 0.934999 0.926999 0.922221
  0.3  0.927498 0.903746 0.891995 0.884994
  0.4  0.904437 0.873322 0.857985 0.848871
  0.5  0.881927 0.843721 0.824963 0.813846
  0.6  0.859964 0.814940 0.792923 0.779911
  0.7  0.838545 0.786973 0.761858 0.747057
  0.8  0.817665 0.759813 0.731759 0.715275
  0.9  0.797320 0.733451 0.702615 0.684551
  1.0  0.777505 0.707878 0.674416 0.654874
  1.1  0.758213 0.683086 0.647148 0.626228
  1.2  0.739438 0.659064 0.620798 0.598598
  1.3  0.721173 0.635800 0.595351 0.571967
  1.4  0.703412 0.613281 0.570793 0.546317
  1.6  0.669366 0.570431 0.524275 0.497882
  1.8  0.637235 0.530404 0.481103 0.453132
  2.0  0.606947 0.493083 0.441128 0.411893
  2.2  0.578427 0.458343 0.404194 0.373984
  2.4  0.551596 0.426057 0.370137 0.339218
  2.6  0.526375 0.396095 0.338793 0.307405
  2.8  0.502682 0.368324 0.309995 0.278355
  3.0  0.480435 0.342614 0.283580 0.251879
  3.2  0.459555 0.318834 0.259385 0.227792
  3.4  0.439962 0.296859 0.237252 0.205915
  3.6  0.421580 0.276565 0.217029 0.186075
  3.8  0.404332 0.257835 0.198571 0.168107
  4.0  0.388148 0.240554 0.181737 0.151855
  4.2  0.372958 0.224615 0.166396 0.137169
  4.4  0.358696 0.209916 0.152424 0.123913
  4.6  0.345301 0.196361 0.139704 0.111956
  4.8  0.332713 0.183860 0.128129 0.101180
  5.0  0.320876 0.172329 0.117597 0.091471
  5.5  0.294240 0.147243 0.095241 0.071228
  6.0  0.271260 0.126669 0.077581 0.055677
  6.5  0.251331 0.109727 0.063604 0.043730
  7.0  0.233948 0.095707 0.052506 0.034541
  7.5  0.218698 0.084039 0.043655 0.027453
  8.0  0.205239 0.074269 0.036560 0.021968
  8.5  0.193294 0.066036 0.030840 0.017702
  9.0  0.182633 0.059053 0.026200 0.014368
  9.5  0.173068 0.053092 0.022411 0.011747
```

10.0 0.164443 0.047971 0.019296 0.009674

Diffusion through a Granulated Media

While there is an analytical solution for diffusion through a sphere of a given radius, there is no closed-form solution for one grain among many, approximating a granulated media. Derivation for the analytical solution listed below for a single grain size is beyond the scope of this book.

$$C(r,t) = \frac{C_S}{2}\left[erfc\left(\frac{r - \frac{Vt}{R} - \frac{L}{2}}{2\sqrt{\frac{Dt}{R}}}\right) - erfc\left(\frac{r - \frac{Vt}{R} + \frac{L}{2}}{2\sqrt{\frac{Dt}{R}}}\right) \right] + \frac{(C_0 - C_S)}{2}\left[erfc\left(\frac{r - \frac{Vt}{R} - \frac{L}{2}}{2\sqrt{\frac{Dt}{R}}}\right) - erfc\left(\frac{r - \frac{Vt}{R} + \frac{L}{2}}{2\sqrt{\frac{Dt}{R}}}\right) \right] \quad (21.3)$$

Numerical integration is used to find a solution for a distribution of grain sizes. We use Gauss Quadrature to integrate from $\mu - \pi\sigma$ to $\mu + \pi\sigma$, where μ is the mean and σ is the standard deviation of the grain size. In order to perform these calculations, it's necessary to pass a group of parameters to the quadrature function. This is done by defining a structure and passing a pointer to it, as illustrated below:

```
typedef struct{
  double Gm;
  double sG;
  double Co;
  double D;
  double r;
  double t;
}DIFU;

double GQuad(double*A,double*W,int n,double
   f(DIFU*,double),DIFU*difu,double a,double b)
{
int i;
double ax,dx,q;
ax=(a+b)/2.;
dx=(b-a)/2.;
for(q=i=0;i<n/2;i++)
   q+=W[i]*(f(difu,ax-dx*A[i])+f(difu,ax+dx*A[i]));
return(q*dx);
}

double Concentration(double Co,double D,double R,double
   r,double t)
```

```c
  {
  double s;
  s=2.*sqrt(D*t)/R;
  t*=(1.131302673065950*s+0.2330227286503510)*s+1.;
  r*=(0.404882980327693*s-0.0539281549540029)*s+1.;
  return(Co*(erfc((r-R)/2./sqrt(D*t))-
    erfc((r+R)/2./sqrt(D*t)))/2.);
  }

double Distribution(double Gm,double sG,double G)
  {
  double g;
  g=(G-Gm)/sG;
  return(exp(-g*g/2.)/sG/sqrt(2.*M_PI));
  }

double Contribution(DIFU*difu,double G)
  {
  return(Distribution(difu->Gm,difu-
    >sG,G)*Concentration(difu->Co,difu-
    >D,pow(10.,G)/20.,difu->r,difu->t));
  }

int main(int argc,char**argv,char**envp)
  {
  double Rm,Rx;
  DIFU difu;
  FILE*fp;
  difu.Co=1.;
  difu.D=1E-5;
  difu.Gm=-0.53;
  difu.sG=0.79;
  Rm=pow(10.,difu.Gm)/20.;
  Rx=Rm*3.;
  fp=fopen("diffusion.csv","wt");
  for(difu.r=0;difu.r<Rx;difu.r+=Rx/100)
     {
     fprintf(fp,"%lG",difu.r);
     if(fabs(difu.r)<=Rm)
       fprintf(fp,",1");
     else
       fprintf(fp,",0");
     for(difu.t=0.001;difu.t<101.;difu.t*=10.)

    fprintf(fp,",%lG",GQuad(A4096,W4096,4096,Contributio
    n,&difu,difu.Gm-difu.sG*M_PI,difu.Gm+difu.sG*M_PI));
     fprintf(fp,"\n");
     }
  fclose(fp);
  return(0);
```

}

All of the associated files may be found in the applications\diffusion folder. The results are shown in the following graph:

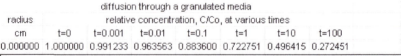

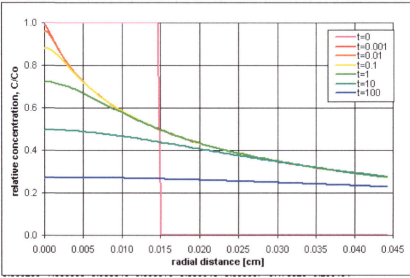

Integral within an Integral

A different solution to this diffusion problem leads to a solution without a closed form, requiring numerical integration. Application of that solution over a range of grain sizes requires another integration, making this an integration within an integration. Again, parameters are passed, but this time instead of creating a type and passing a pointer, the parameters are simply passed through the integrator functions as additional arguments. The code (source.c) and associated files may be found in the applications\diffusion folder.

```
double Spherical(double Co,double D,double R,double
   r,double t)
{
double s;
if(t<FLT_EPSILON)
   {
   if(r<-R)
     return(0.);
   if(r>R)
     return(0.);
   return(Co);
   }
```

```
  s=2*sqrt(D*t)/R;
  t*=(1.131302673065950*s+0.2330227286503510)*s+1.;
  r*=(0.404882980327693*s-0.0539281549540029)*s+1.;
  return(Co*(erfc((r-R)/2./sqrt(D*t))-
    erfc((r+R)/2./sqrt(D*t)))/2.);
  }
double OneGrain(double R,double Z,double X,double t)
  {
  return(Spherical(Co,D,R,R,t/Retardation-(Z-L-X)/U)
        -Spherical(Co,D,R,R,t-(Z-L-X)/U*Retardation));
  }
double DistributedGrains(double Z,double X,double t)
  {
  int i;
  double aG,dG,g,G,G1,G2,P,Q,R;
  G1=Gm-sG*M_PI;
  G2=Gm+sG*M_PI;
  dG=(G2-G1)/2.;
  aG=(G1+G2)/2.;
  for(Q=i=0;i<48;i++)
    {
    G=aG-dG*A96[i];
    R=pow(10.,G)/20.;
    g=(G-Gm)/sG;
    P=exp(-g*g/2.)/sG/sqrt(2.*M_PI);
    Q+=P*W96[i]*OneGrain(R,Z,X,t);
    G=aG+dG*A96[i];
    R=pow(10.,G)/20.;
    g=(G-Gm)/sG;
    P=exp(-g*g/2.)/sG/sqrt(2.*M_PI);
    Q+=P*W96[i]*OneGrain(R,Z,X,t);
    }
  return(Q*dG);
  }
double IntegrateOverSourceZone(double Z,double t)
  {
  int i;
  double Q;
  for(Q=i=0;i<48;i++)
    {
    Q+=W96[i]*DistributedGrains(Z,(1.-A96[i])/2.,t);
    Q+=W96[i]*DistributedGrains(Z,(1.+A96[i])/2.,t);
    }
  return(Q/2);
  }
```

The double integral is utilized as in the following for three different distances:

```
IntegrateOverSourceZone(40.*2.54*12.,t)
IntegrateOverSourceZone(48.*2.54*12.,t)
```

```
IntegrateOverSourceZone(50.*2.54*12.,t)
```
The resulting concentration profiles are shown in the figure below:

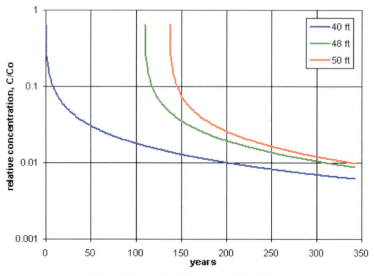

Hybrid Analytical/Numerical Solution

One of the critical outcomes of contaminant transport analyses is breakthrough time, that is, answering the question, "When will the schmutz escape from the containment barrier or when will it arrive at a certain location?" The classical approach (pure diffusive spreading) greatly underestimates this time, as does the zero (i.e., infinitesimal) grain size solution. The finite grain size solution has no closed form and requires integration:

```
double Slab( /* diffusion through a slab */
  double D, /* diffusion coefficient [ftý/year] */
  double L, /* thickness of slab [ft] */
  double X, /* distance from center of slab [ft] */
  double T) /* time [yr] */
  {
  if(T<FLT_EPSILON)
    {
    if(X<-L/2.)
      return(0.);
    if(X>L/2.)
      return(0.);
    return(1.);
    }
  return((erfc((X-L/2.)/2./sqrt(D*T))
        -erfc((X+L/2.)/2./sqrt(D*T)))/2.);
  }
```

```
double Sphere(/* diffusion through a sphere */
  double D, /* diffusion coefficient [ftý/year] */
  double G, /* radius of grain [ft] */
  double R, /* radial distance from center of grain
    [ft] */
  double T) /* time [yr] */
  {
  double S;
  if(T<FLT_EPSILON)
    {
    if(R<-G)
      return(0.);
    if(R>G)
      return(0.);
    return(1.);
    }
  S=2.*sqrt(D*T)/G;
  T*=(1.131302673065950*S+0.2330227286503510)*S+1.;
  R*=(0.404882980327693*S-0.0539281549540029)*S+1.;
  return((erfc((R-G)/2./sqrt(D*T))
         -erfc((R+G)/2./sqrt(D*T)))/2.);
  }
double Lumped(/* lumped system model */
  double Dh, /* diffusion coefficient [ftý/year] */
  double Ra, /* advective retardation coefficient */
  double Rd, /* diffusive retardation coefficient */
  double Vs, /* seepage velocity [ft/yr] */
  double Lc, /* contaminated length [ft] */
  double X, /* distance from center of contamination
    [ft] */
  double T) /* time [yr] */
  {
  X-=Vs*T/Ra;
  return(Slab(Dh,Lc,X,T/Rd));
  }
double Granular(/* single-grain spherical model */
  double Dh, /* diffusion coefficient [ftý/year] */
  double Ra, /* advective retardation coefficient */
  double Rg, /* radius of grain [ft] */
  double Vs, /* seepage velocity [ft/yr] */
  double Lc, /* contaminated length [ft] */
  double X, /* distance from center of contamination
    [ft] */
  double T) /* time [yr] */
  {
  double Rm,S;
  X-=Vs*T/Ra;
  Rm=sqrt(Rg*Lc/2.);
  S=2.*M_PI*sqrt(Dh*T/Ra)/Rm;
  return(Slab(Dh,Lc,X,T/Ra)*exp(-S*S));
```

```c
  }
double Distributed(/* multi-grain spherical model */
  double Dh, /* diffusion coefficient [ftý/year] */
  double Ra, /* advective retardation coefficient */
  double Gm, /* log10(mean radius of grain [ft]) */
  double sG, /* standard deviation of Gm */
  double Vs, /* seepage velocity [ft/yr] */
  double Lc, /* contaminated length [ft] */
  double X, /* distance from center of contamination
    [ft] */
  double T) /* time [yr] */
  {
  int i;
  double aG,dG,g,G,G1,G2,P,Q,Rg;
  G1=Gm-sG*M_PI;
  G2=Gm+sG*M_PI;
  dG=(G2-G1)/2.;
  aG=(G1+G2)/2.;
  for(Q=i=0;i<48;i++)
    {
    G=aG-dG*A96[i];
    Rg=pow(10.,G);
    g=(G-Gm)/sG;
    P=exp(-g*g/2)/sG/sqrt(2.*M_PI);
    Q+=P*W96[i]*Granular(Dh,Ra,Rg,Vs,Lc,X,T);
    G=aG+dG*A96[i];
    Rg=pow(10.,G);
    g=(G-Gm)/sG;
    P=exp(-g*g/2.)/sG/sqrt(2.*M_PI);
    Q+=P*W96[i]*Granular(Dh,Ra,Rg,Vs,Lc,X,T);
    }
  return(Q*dG);
  }
double Classical(double x,double t)
  {
  return(Cp/2*erfc((x-Vs*t/Ra)/2/sqrt(Da*t/Ra)));
  }
double InfiniteGrain(double x,double t)
  {
  if(Co>Cs)
    return(Cs*Lumped(Da,Ra,1,Vs,Lc,x+Lc/2,t)
      +(Co-Cs)*Lumped(Dd,Rd,1,Vs,Lc,x+Lc/2,t));
  else
    return(Co*Lumped(Da,Ra,1,Vs,Lc,x+Lc/2,t));
  }
double FiniteGrain(double x,double t)
  {
  if(Co>Cs)
    return(Cs*Distributed(Da,Ra,Gm,sG,Vs,Lc,x+Lc/2,t)
      +(Co-Cs)*Distributed(Dd,Rd,Gm,sG,Vs,Lc,x+Lc/2,t));
```

```
  else
    return(Co*Distributed(Da,Ra,Gm,sG,Vs,Lc,x+Lc/2,t));
}
```

The code (breakthrough.c) and associated files may also be found in the applications\diffusion folder. The breakthrough curves are shown in the following figure:

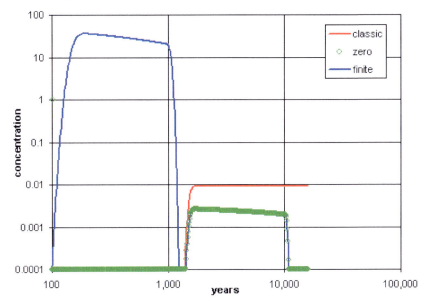

The zero grain size solution (green circles) matches the second part of the finite grain size solution (blue curve), but not the first, which is a lot bigger and a lot sooner! The classic approach (red curve) misses the mark by a mile.

Boundary Element Method

The boundary element method uses Green's Lemma to transform an area integral into a boundary one. Results of numerical boundary integrations go into a matrix solver (simultaneous linear equations) that produces solutions to partial differential equations (PDEs) over an irregular domain. The most common PDE solved in this manner is Laplace's[28], which appears in field theory (gravity, magnetism), invicid fluid flow, and steady-state heat conduction.

There are 3 different integrals: 1) along a boundary adjacent to the current node, 2) along a boundary not directly connected to the current node, and 3)

---

[28] Pierre-Simon the marquis de Laplace (1749-1847): French mathematician, physicist, and astronomer.

inside the boundary. These 3 functions are all integrated using the same 96-pt Gauss Quadrature:

```
void AdjacentBoundaryIntegral(double X1,double
   Y1,double X2,double Y2,double*G1,double*G2)
  {
  double A,dS;
  dS=hypot(X2-X1,Y2-Y1);
  if(dS<DBL_EPSILON)
    Abort(__LINE__,"adjacent boundary integral
    encountered coincident points");
  A=log(dS);
  *G1=dS*(1.5-A)/2.;
  *G2=dS*(0.5-A)/2.;
  }
void RemoteBoundaryIntegral(double Xp,double Yp,double
   X1,double Y1,double X2,double
   Y2,double*H1,double*H2,double*G1,double*G2)
  {
  int i,n;
  double
   Ai,aX,aY,bX,bY,D,dS,H21,H22,G21,G22,P1,P2,Q,R,T,U,Wi
   ,X,Y;
  bX=(X2+X1)/2.;
  bY=(Y2+Y1)/2.;
  aX=(X2-X1)/2.;
  aY=(Y2-Y1)/2.;
  dS=hypot(aX,aY);
  if(fabs(aX)>dS/1000.)
     {
     T=aY/aX;
     D=fabs(T*Xp-Yp+Y1-T*X1)/sqrt(T*T+1.);
     }
  else
     D=fabs(Xp-X1);
  if((X1-Xp)*(Y2-Yp)-(X2-Xp)*(Y1-Yp)<0.)
     D=-D;
  H21=0.;
  H22=0.;
  G21=0.;
  G22=0.;
  n=96;
  for(i=0;i<n;i++)
     {
     if(i<n/2)
        {
        Wi= W96[n/2-1-i];
        Ai=-A96[n/2-1-i];
        }
     else
        {
```

```
      Wi= W96[i-n/2];
      Ai= A96[i-n/2];
      }
    P1=(1.-Ai)/2.;
    P2=(1.+Ai)/2.;
    X=bX+aX*Ai;
    Y=bY+aY*Ai;
    R=hypot(X-Xp,Y-Yp);
    if(R<DBL_EPSILON)
      Abort(__LINE__,"remote boundary integral
    encountered coincident points");
    U=-log(R);
    Q=-D/(R*R);
    H21+=Wi*P1*Q;
    H22+=Wi*P2*Q;
    G21+=Wi*P1*U;
    G22+=Wi*P2*U;
    }
  *H1=H21*dS;
  *H2=H22*dS;
  *G1=G21*dS;
  *G2=G22*dS;
  }
int InternalIntegral(double Xp,double Yp,double
    X1,double Y1,double X2,double Y2,
double*H1o,double*H2o,double*G1o,double*G2o,double*H1x,
    double*H2x,
double*H1y,double*H2y,double*G1x,double*G2x,double*G1y,
    double*G2y)
  {
  int i,n;
  double
    Ai,aX,aY,bX,bY,D,dS,dX,dY,P1,P2,Q,Qx,Qy,R,Ri,S,T,U,U
    x,Uy,Wi,X,Xi,Y,Yi;
  double H21o,H22o;
  double H21x,H22x;
  double H21y,H22y;
  double G21o,G22o;
  double G21x,G22x;
  double G21y,G22y;
  aX=(X2-X1)/2.;
  aY=(Y2-Y1)/2.;
  bX=(X2+X1)/2.;
  bY=(Y2+Y1)/2.;
  dS=hypot(aX,aY);
  if(fabs(aX)>dS/1000.)
    {
    T=aY/aX;
    S=sqrt(T*T+1.);
    D=(T*Xp-Yp+Y1-T*X1)/S;
```

```
   if(D>=0.)
      {
      dX=T/S;
      dY=-1./S;
      }
   else
      {
      D=-D;
      dX=-T/S;
      dY=1./S;
      }
   }
else
   {
   D=Xp-X1;
   dY=0.;
   if(D>=0.)
      dX=1.;
   else
      {
      D=-D;
      dX=-1.;
      }
   }
if((X1-Xp)*(Y2-Yp)<(X2-Xp)*(Y1-Yp))
   {
   D=-D;
   dX=-dX;
   dY=-dY;
   }
H21o=H22o=H21x=H22x=H21y=H22y=0.;
G21o=G22o=G21x=G22x=G21y=G22y=0.;
n=96;
for(i=0;i<n;i++)
   {
   if(i<n/2)
      {
      Wi= W96[n/2-1-i];
      Ai=-A96[n/2-1-i];
      }
   else
      {
      Wi= W96[i-n/2];
      Ai= A96[i-n/2];
      }
   P1=(1.-Ai)/2.;
   P2=(1.+Ai)/2.;
   X=bX+aX*Ai;
   Y=bY+aY*Ai;
   Xi=X-Xp;
```

```
    Yi=Y-Yp;
    Ri=Xi*Xi+Yi*Yi;
    if(Ri<An/1000.)
       return(1);
    R=sqrt(Ri);
    U=-log(R);
    Ux=Xi/Ri;
    Uy=Yi/Ri;
    Q=-D/(R*R);
    Qx=(-2.*D*Xi/Ri-dX)/Ri;
    Qy=(-2.*D*Yi/Ri-dY)/Ri;
    H21o+=Wi*P1*Q;
    H22o+=Wi*P2*Q;
    H21x+=Wi*P1*Qx;
    H22x+=Wi*P2*Qx;
    H21y+=Wi*P1*Qy;
    H22y+=Wi*P2*Qy;
    G21o+=Wi*P1*U;
    G22o+=Wi*P2*U;
    G21x+=Wi*P1*Ux;
    G22x+=Wi*P2*Ux;
    G21y+=Wi*P1*Uy;
    G22y+=Wi*P2*Uy;
    }
 *H1o=H21o*dS;
 *H2o=H22o*dS;
 *H1x=H21x*dS;
 *H2x=H22x*dS;
 *H1y=H21y*dS;
 *H2y=H22y*dS;
 *G1o=G21o*dS;
 *G2o=G22o*dS;
 *G1x=G21x*dS;
 *G2x=G22x*dS;
 *G1y=G21y*dS;
 *G2y=G22y*dS;
 return(0);
 }
```

The first two integrals are needed to set up the problem and solve for the potential at the boundary points; whereas, the third is needed to evaluate the potential inside the boundary. The results are quite interesting. This first example is for a river bend.

Here are the velocity vectors:

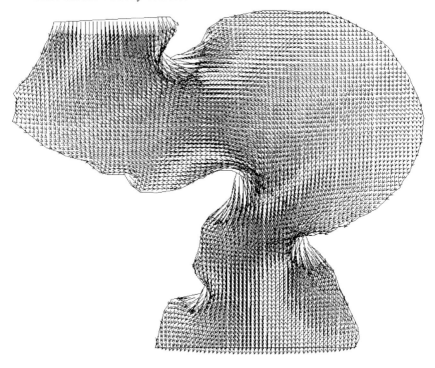

This next figure shows the potential or stream lines:

This next example is for flow in a lake around two islands. First, the velocity vectors:

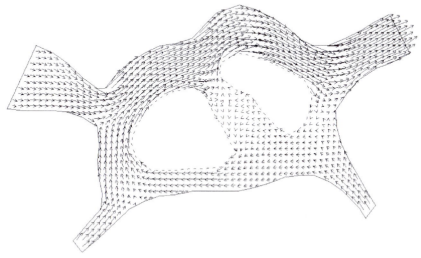

Then the streamlines:

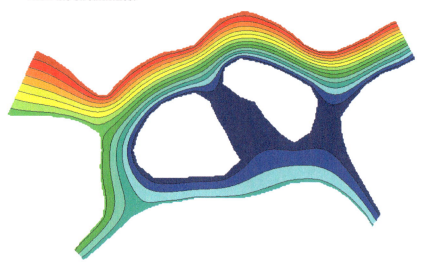

This third example is flow through a natural draft cooling tower:

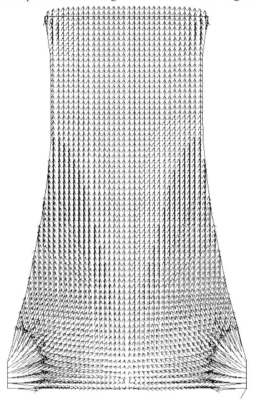

The code (pflow.c), input files, and output are in the applications\pflow folder in the on-line archive. This is a very useful tool and quick to set up. The results are displayed with my general-purpose 2D/3D graphics program, TP2, which is also available free at the same web site.

Heat Exchangers

Heat exchangers are sized according to how demanding the requirements are for the transfer of heat—not just how much heat is transferred, but how close the hot and cold side temperatures approach. A *close approach* heat exchanger is much more expensive than a *long approach* one because it takes that much more material to construct and effort to fabricate. In the case of constant thermal properties, the heat transfer problem has a closed form solution: the log-mean temperature difference. Solutions with variable properties are beyond the scope of this book and readers are directed to my work on that subject:

http://www.amazon.com/dp/B078HM16GV

Within the context of numerical calculus we will only consider constant properties. A spreadsheet (heater.xls) is provided in the applications\heater folder that illustrates the analytical (closed-form) solution along with two numerical ones: 5-pt Lobatto and 4-pt Chebyshev. In this case, the closed-form solution is exact and the numerical ones are approximate. The results are shown in the following figure for typical inputs:

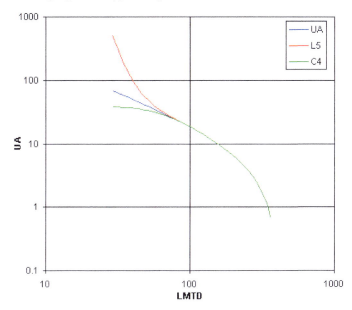

Note that the two numerical solutions (red and green curves) diverge and in opposite directions compared to the analytical solution (blue curve). The 5-pt Lobatto solution (red curve) contains the end points and diverges high (above the analytical) and the 4-pt Chebyshev solution (green curve) does not contain the end points and diverges low (below the analytical). This is important to know when evaluating the integral with variable properties and there is no analytical solution. Some improvement can be gained by averaging the two numerical solutions in the case of variable properties, although using a higher order method yields a superior result in every case that I have ever been able to check.

Cooling Tower Demand Curves

The derivation of cooling tower demand curves is beyond the scope of this book, but is covered thoroughly in Reference 11. Originally, Merkel chose to use the 4-pt Chebyshev method, which is an open method and doesn't include the end points. This works adequately for many problems, but not for *close approach* or high performance designs. An evaporative cooling tower is a type of heat exchanger and the concept of *pinch points* applies here as well. For this type of system, pinches will always occur at the air inlet/water outlet.

The following code calculates the demand using the Merkel (4-pt Chebyshev) and the Exact (10-pt Lobatto) methods:

```
double KMerkel(double Twb,double Ran,double App,double
  LG)
{
int i;
double dHa,Ha,Hai,Hw,KaV,Tco,Tw;
static double a[4]={0.10,0.40,0.60,0.90};
static double w[4]={0.25,0.25,0.25,0.25};
Tco=Twb+App;
Hai=fHtwb(user.baro,Twb);
dHa=Ran*LG;
KaV=0;
for(i=0;i<4;i++)
  {
  Ha=Hai+a[i]*dHa;
  Tw=Tco+a[i]*Ran;
  Hw=fHtwb(user.baro,Tw);
  if(Hw<=Ha)
    return(999.9);
  KaV=KaV+w[i]/(Hw-Ha);
  }
KaV=Ran*KaV;
return(KaV);
}
double KExact(double Twb,double Ran,double App,double
  LG)
{
int i;
double
  Ha,Hai,Hao,KaV,Ta,Tai,Tao,Tao1,Tao2,Tco,Tho,Tw,W,Wai
  ,Wao,Ws;
static double a[10]={0.000000,0.040233,0.130613,
  0.261038,0.417361,0.582639,0.738962,0.869387,
  0.959767,1.00000};
static double w[10]={0.011111,0.066653,0.112445,
  0.146021,0.163770,0.163770,0.146021,0.112445,
  0.066653,0.011111};
Tco=Twb+App;
Tho=Tco+Ran;
Tai=Twb;
Hai=fHtwb(user.baro,Tai);
Wai=fWsat(user.baro,Tai);
Tao1=Twb;
Tao2=Tho;
for(i=0;i<32;i++)
  {
  Tao=(Tao1+Tao2)/2.;
  Hao=fHtwb(user.baro,Tao);
  Wao=fWsat(user.baro,Tao);
```

```
    if(Hao<Hai+(LG*Ran+(Wao-Wai)*(Tco-32.018)))
       Tao1=Tao;
    else
       Tao2=Tao;
    }
 KaV=0;
 for(i=0;i<10;i++)
    {
    Ha=Hao+a[i]*(Hai-Hao);
    Ta=fTwbh(user.baro,Ha);
    W=fWsat(user.baro,Ta);
    Tw=((LG*Tho-(Wao-W)*32.018)-Hao+Ha)/(LG-Wao+W);
    Ws=fWsat(user.baro,Tw);
    if(Tw<=Ta)
       return(999.9);
    KaV=KaV+w[i]/(Lewis*0.2406*(Tw-Ta)
    +(1061.39+0.427933*Ta)*(Ws-W)/(1+W));
    }
 KaV=Ran*KaV;
 return(KaV);
 }
```

I have provided the entire code, not only these calculations, but also the Windows® interface as well. You will find all of the files, resources, and compiler commands (_compile.bat) in the applications\KaVL folder. For clarity of comparison, the abscissas for both the Chebyshev and Lobatto integrations have been modified so that they work over the interval 0 to 1 instead of -1 to +1.

The are some additional features in this program you might find useful, including log-linear regression model and associated statistics plus copying data or a graph onto the clipboard. The program will plot performance data (in red) on top of the demand curves (in black). This is how you design a cooling tower: match supply and demand. Supply here is the cooling capacity of the packing (plastic stuff) and demand is the cooling capacity. Where the two meet (i.e., red crosses black) is where the cooling tower will operate. I have included data files for 24 common types of packing (*.?FT). The beginning of the file name is the packing type and the end is the depth in feet. This is the way these devices are typically designed. Ordinarily, you would design a cooling tower to operate in the middle of the middle of the red circles.

The program draws the curves on the unusual green log-log graph paper format that has been used for many decades in the cooling tower industry. Run the program, specify an operating condition and formulation plus optional data file and press the OK button. Select File/Export to copy the graph or numbers to the clipboard. Ctl-V will paste it into any Microsoft® Office® document, as illustrated in this next figure:

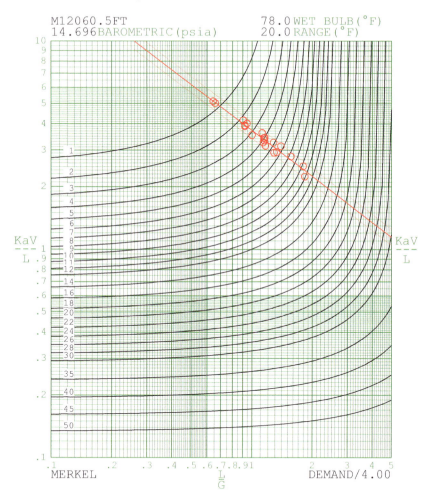

The program works in both English and SI units. The preceding graph is dimensionless but the temperature inputs have dimensions. The results are slightly different for the Exact method using 10-pt Lobatto, as illustrated in this next figure:

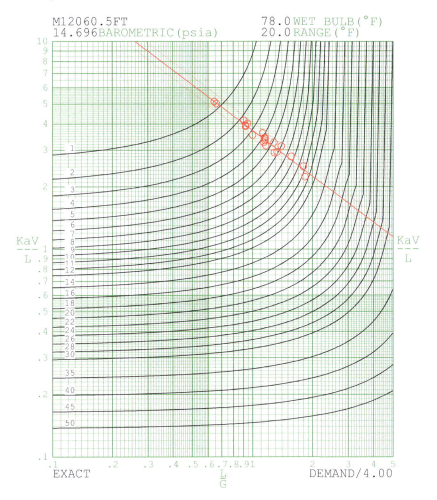

For these operating conditions, the two calculations are not much different. This next figure compares the two integration methods for the most extreme case (1° approach), which would not be a practical design:

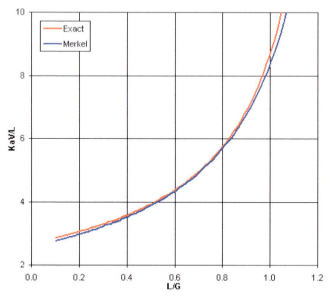

A topic that's not covered in this book—partial differential equations—is included in this example. Merkel's equation only works for counterflow cooling towers. It doesn't work for crossflow. Cooling towers are designed with a crossflow arrangement for a variety of reasons, including the avoidance of fouling or plugging. This program solves the crossflow partial differential equation in 2D using the 4-th order Runge-Kutta method. The program also includes moist air properties, which are essential to the calculations.

Laplace Transforms

The Laplace transform is defined by the following integral:

$$F(s) = \int_0^\infty e^{-st} f(t) dt \qquad (21.4)$$

Tables abound for such transforms having a closed-form solution; however, there are many more functions for which no analytical solution has been found. In such cases, it is possible to evaluate this integral numerically. A simple modification of the first two of four formulas used in Chapter 20 will perform this special improper integral.

```
double GQ0Ia(double*A,double*W,int n,double
   f(int,double),int m,double s)
  {
  int i;
  double q,t,x;
  for(q=i=0;i<n/2;i++)
    {
    x=(1.-A[i])/2.;
    t=1./x-1.;
    q+=W[i]*exp(-s*t)*f(m,t)/x/x;
    x=(1.+A[i])/2.;
    t=1./x-1.;
    q+=W[i]*exp(-s*t)*f(m,t)/x/x;
    }
  return(q/2.);
  }
double GQ0Ib(double*A,double*W,int n,double
   f(int,double),int m,double s)
  {
  int i;
  double q,t,x;
  for(q=i=0;i<n/2;i++)
    {
    x=(1.-A[i])/2.;
    t=-log(x);
    q+=W[i]*exp(-s*t)*f(m,t)/x;
    x=(1.+A[i])/2.;
    t=-log(x);
    q+=W[i]*exp(-s*t)*f(m,t)/x;
    }
  return(q/2.);
  }
```

To check the implementation, we will first evaluate simple integrals for which there is a known solution, specifically:

$$L\{t^m\} = \frac{\Gamma(m+1)}{s^{m+1}} \qquad (21.5)$$

Typical program output is listed below:
```
Laplace Transform by Gauss Quadrature
m  s    result1  result2  exact
1 0.5   4.00000  3.99552  4.00000
1 1.0   1.00000  1.00000  1.00000
1 1.5   0.44444  0.44444  0.44444
1 2.0   0.25000  0.25000  0.25000
1 2.5   0.16000  0.16000  0.16000
1 3.0   0.11111  0.11111  0.11111
1 3.5   0.08163  0.08163  0.08163
1 4.0   0.06250  0.06250  0.06250
1 4.5   0.04938  0.04938  0.04938
```

```
1 5.0  0.04000   0.04000   0.04000
2 0.5  16.00000  15.90589  16.00000
2 1.0  2.00000   2.00000   2.00000
2 1.5  0.59259   0.59259   0.59259
2 2.0  0.25000   0.25000   0.25000
2 2.5  0.12800   0.12800   0.12800
2 3.0  0.07407   0.07407   0.07407
2 3.5  0.04665   0.04665   0.04665
2 4.0  0.03125   0.03125   0.03125
2 4.5  0.02195   0.02195   0.02195
2 5.0  0.01600   0.01600   0.01600
3 0.5  96.00000  94.01740  96.00000
3 1.0  6.00000   5.99996   6.00000
3 1.5  1.18519   1.18519   1.18519
3 2.0  0.37500   0.37500   0.37500
3 2.5  0.15360   0.15360   0.15360
3 3.0  0.07407   0.07407   0.07407
3 3.5  0.03998   0.03998   0.03998
3 4.0  0.02344   0.02344   0.02344
3 4.5  0.01463   0.01463   0.01463
3 5.0  0.00960   0.00960   0.00960
4 1.0  24.00000  23.99904  24.00000
4 1.5  3.16049   3.16049   3.16049
4 2.0  0.75000   0.75000   0.75000
4 2.5  0.24576   0.24576   0.24576
4 3.0  0.09877   0.09877   0.09877
4 3.5  0.04570   0.04570   0.04570
4 4.0  0.02344   0.02344   0.02344
4 4.5  0.01301   0.01301   0.01301
4 5.0  0.00768   0.00768   0.00768
```

The numerical results are quite accurate even up to the 4th power. A Laplace Transform that arises in molecular dynamics is the following function:

$$f(t) = t^n \, erfc\left(\frac{\sqrt{t}}{2}\right) \qquad (21.6)$$

The error function and its complement can be approximated by:

```
double erf(double x)
  {
  double e,q,t,y;
  if(fabs(x)<FLT_EPSILON)
    return(0.);
  if(x>6.)
    return(1.);
  y=fabs(x);
  t=1./(1.+0.3275911*y);
  q=((((10.061405429*t-10.453152027)*t+10.421413741)*t-
    0.284496736)*t+0.254829592)*t;
  e=1.-q/exp(x*x);
```

```
  if(x<0)
    return(-e);
  return(e);
  }
#define erfc(x) (1.-erf(x))
double f(int m,double t)
  {
  return(pow(t,m)*erfc(sqrt(t)/2.));
  }
```

The results are difficult to verify, but reasonable:

```
Laplace Transform by Gauss Quadrature
m  s     result1   result2
0  0.5   7.80140   7.80140
0  1.0   5.19106   5.19106
0  1.5   3.93109   3.93109
0  2.0   3.17930   3.17930
0  2.5   2.67662   2.67662
0  3.0   2.31544   2.31544
0  3.5   2.04272   2.04272
0  4.0   1.82913   1.82913
0  4.5   1.65711   1.65711
0  5.0   1.51546   1.51546
1  0.5   8.10293   8.10290
1  1.0   3.38576   3.38576
1  1.5   1.88014   1.88014
1  2.0   1.20388   1.20388
1  2.5   0.84020   0.84020
1  3.0   0.62132   0.62132
1  3.5   0.47899   0.47899
1  4.0   0.38106   0.38106
1  4.5   0.31069   0.31069
1  5.0   0.25837   0.25837
2  0.5  18.18250  18.18180
2  1.0   4.69377   4.69377
2  1.5   1.89431   1.89431
2  2.0   0.95471   0.95471
2  2.5   0.55002   0.55002
2  3.0   0.34657   0.34657
2  3.5   0.23287   0.23287
2  4.0   0.16424   0.16424
2  4.5   0.12030   0.12030
2  5.0   0.09083   0.09083
3  0.5  63.57390  63.55861
3  1.0  10.06008  10.06008
3  1.5   2.93785   2.93785
3  2.0   1.16206   1.16206
3  2.5   0.55148   0.55148
3  3.0   0.29562   0.29562
3  3.5   0.17291   0.17291
```

```
3 4.0  0.10800  0.10800
3 4.5  0.07100  0.07100
3 5.0  0.04864  0.04864
4 1.0 29.28820 29.28820
4 1.5  6.17284  6.17284
4 2.0  1.91294  1.91294
4 2.5  0.74689  0.74689
4 3.0  0.34026  0.34026
4 3.5  0.17311  0.17311
4 4.0  0.09570  0.09570
4 4.5  0.05644  0.05644
4 5.0  0.03506  0.03506
```

Both of these programs may be found in the applications\Laplace folder in the on-line archive.

Gamma and Psi Functions

The gamma and digamma (or psi) functions are defined in terms of integrals and also have known infinite series solutions as well as accurate approximations. The gamma function is defined by the integral:

$$\Gamma(z) = \int_0^\infty t^{z-1} e^{-t} dt \qquad (21.7)$$

The gamma function provides another test of improper integral evaluation using Gauss Quadrature. The same two functions are modified to handle this special case:

```
double GQ0Ia(double*A,double*W,int n,double
   f(double,double),double z)
   {
   int i;
   double q,t,x;
   for(q=i=0;i<n/2;i++)
     {
     x=(1.-A[i])/2.;
     t=1./x-1.;
     q+=W[i]*f(z,t)/x/x;
     x=(1.+A[i])/2.;
     t=1./x-1.;
     q+=W[i]*f(z,t)/x/x;
     }
   return(q/2.);
   }
double GQ0Ib(double*A,double*W,int n,double
   f(double,double),double z)
   {
   int i;
   double q,t,x;
   for(q=i=0;i<n/2;i++)
     {
```

```
   x=(1.-A[i])/2.;
   t=-log(x);
   q+=W[i]*f(z,t)/x;
   x=(1.+A[i])/2.;
   t=-log(x);
   q+=W[i]*f(z,t)/x;
   }
  return(q/2.);
  }
double f(double z,double t)
  {
  return(pow(t,z-1.)*exp(-t));
  }
```

The approximate solution is also simple:
```
double gamma(double x)
  {
  static double c[9]={1.,-0.577191652,0.988205891,-
    0.897056937,0.918206857,
    -0.756704078,0.482199394,-0.193527818,0.035868343};
  int i;
  double g,y,z;
  y=x;
  g=1.;
  while(y>2.)
     {
     y-=1.;
     g*=y;
     }
  while(y<1.)
     {
     g/=y;
     y+=1.;
     }
  z=c[8];
  for(i=7;i>=0;i--)
     {
     z*=y-1.;
     z+=c[i];
     }
  return(z*g);
  }
```

These constants may be found in Abramowitz and Stegun. The recurrence relationship is used to shift the value into the range 1 to 2, where this approximation is more than adequate. The quadrature results are quite accurate:

```
Gamma Function by Gauss Quadrature
 z    result1   result2   exact
0.5  1.772241  1.772241  1.772454
0.6  1.489167  1.489167  1.489193
0.7  1.298052  1.298052  1.298055
```

```
0.8 1.164229 1.164229 1.164230
0.9 1.068629 1.068629 1.068629
1.0 1.000000 1.000000 1.000000
1.1 0.951351 0.951351 0.951351
1.2 0.918169 0.918169 0.918169
1.3 0.897471 0.897471 0.897471
1.4 0.887264 0.887264 0.887264
1.5 0.886227 0.886227 0.886227
1.6 0.893515 0.893515 0.893516
1.7 0.908639 0.908639 0.908639
1.8 0.931384 0.931384 0.931384
1.9 0.961766 0.961766 0.961766
2.0 1.000000 1.000000 1.000000
2.1 1.046486 1.046486 1.046486
2.2 1.101802 1.101802 1.101803
2.3 1.166712 1.166712 1.166712
2.4 1.242169 1.242169 1.242169
2.5 1.329340 1.329340 1.329340
```

The digamma (or Psi) function is given by:

$$\psi(z) = \frac{d \ln[\Gamma(z)]}{dz} \tag{21.8}$$

This derivative is also equal to the following integral:

$$\psi(z) = -\gamma + \int_0^1 \left(\frac{1 - t^{z-1}}{1 - t} \right) dt \tag{21.9}$$

This integral (Equation 21.9) can also be calculated using Gauss Quadrature using the methods outlined in Chapter 24. The differential (Equation 21.8) can also easily be calculated using the gamma function listed above, as shown below:

```
double digamma(double z)
{
double dz;
dz=0.0001;
return((gamma(z+dz/2.)-gamma(z-dz/2.))/dz/gamma(z));
}
```

The results are quite accurate:

```
digamma (Psi) function
  z    integra  differn
0.5  -1.96330 -1.96351
0.6  -1.54059 -1.54062
0.7  -1.22002 -1.22003
0.8  -0.96501 -0.96500
0.9  -0.75493 -0.75493
1.0  -0.57722 -0.57719
1.1  -0.42375 -0.42376
```

```
1.2  -0.28904  -0.28904
1.3  -0.16919  -0.16920
1.4  -0.06138  -0.06138
1.5   0.03649   0.03649
1.6   0.12605   0.12605
1.7   0.20855   0.20854
1.8   0.28499   0.28500
1.9   0.35618   0.35618
2.0   0.42278   0.42281
2.1   0.48534   0.48534
2.2   0.54429   0.54430
2.3   0.60004   0.60004
2.4   0.65290   0.65290
2.5   0.70316   0.70316
```

Both programs (gamma.c and digamma.c), along with a batch file to compile them, may be found in the applications\gamma folder.

Incomplete Gamma & Beta Functions

The incomplete gamma and beta functions appear in various derivations throughout applied mathematics. Although these can be calculated from infinite series, the coefficients are complicated and many terms are often required. These can also be estimated numerically by quadrature. The incomplete gamma function is defined by the following two integral:

$$\Gamma_x(a) = \int_0^x e^{-t} t^{a-1} dt \qquad (21.10)$$

The previous functions can be adapted as follows:

```
double GQ(double*A,double*W,int n,double
   f(double,double),double a,double x)
  {
  int i;
  double q,t;
  for(q=i=0;i<n/2;i++)
    {
    t=x*(1.-A[i])/2.;
    q+=W[i]*f(a,t);
    t=x*(1.+A[i])/2.;
    q+=W[i]*f(a,t);
    }
  return(x*q/2.);
  }

double f(double a,double t)
  {
  return(pow(t,a-1.)*exp(-t));
  }
double gstar(double a,double x)
  {
```

```
    return(pow(x,-
      a)*GQ(A4096,W4096,4096,f,a,x)/gamma(a));
    }
```

The output of the program (igamma.c) is listed below:

```
incomplete gamma function
  a    x    g*(a,x)
0.5  0.5  0.965349
0.5  1.0  0.842581
0.5  1.5  0.748391
0.5  2.0  0.674813
0.5  2.5  0.616305
0.5  3.0  0.568971
1.0  0.5  0.786939
1.0  1.0  0.632121
1.0  1.5  0.517913
1.0  2.0  0.432332
1.0  2.5  0.367166
1.0  3.0  0.316738
1.5  0.5  0.562144
1.5  1.0  0.427593
1.5  1.5  0.331157
1.5  2.0  0.261112
1.5  2.5  0.209521
1.5  3.0  0.170971
2.0  0.5  0.360816
2.0  1.0  0.264241
2.0  1.5  0.196522
2.0  2.0  0.148499
2.0  2.5  0.114032
2.0  3.0  0.088984
2.5  0.5  0.211760
2.5  1.0  0.150855
2.5  1.5  0.108871
2.5  2.0  0.079653
2.5  2.5  0.059109
2.5  3.0  0.044506
3.0  0.5  0.115101
3.0  1.0  0.080301
3.0  1.5  0.056638
3.0  2.0  0.040415
3.0  2.5  0.029196
3.0  3.0  0.021363
```

A map can be drawn using the graphics program mentioned previously:

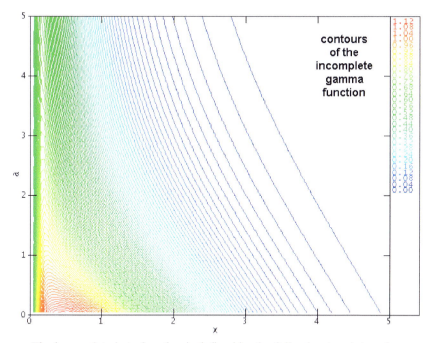

contours of the incomplete gamma function

The incomplete beta function is defined by the following two integral:

$$B_x(a,b) = \int_0^x t^{a-1}(1-t)^{b-1} dt \qquad (21.11)$$

The same functions can be adapted as follows:

```
double GQ(double*A,double*W,int n,double
   f(double,double,double),double a,double b,double x)
   {
   int i;
   double q,t;
   for(q=i=0;i<n/2;i++)
      {
      t=x*(1.-A[i])/2.;
      q+=W[i]*f(a,b,t);
      t=x*(1.+A[i])/2.;
      q+=W[i]*f(a,b,t);
      }
   return(x*q/2.);
   }
double f(double a,double b,double t)
   {
   return(pow(t,a-1.)*pow(1.-t,b-1.));
   }
double ibeta(double a,double b,double x)
   {
```

```
    return(GQ(A4096,W4096,4096,f,a,b,x));
  }
```

The results are as follows:

```
incomplete beta function
  x    a    b   beta(a,b,x)
 0.2  0.5  0.5  0.927200
 0.2  0.5  1.0  0.894332
 0.2  0.5  1.5  0.863553
 0.2  0.5  2.0  0.834704
 0.2  1.0  0.5  0.211146
 0.2  1.0  1.0  0.200000
 0.2  1.0  1.5  0.189639
 0.2  1.0  2.0  0.180000
 0.2  1.5  0.5  0.063648
 0.2  1.5  1.0  0.059628
 0.2  1.5  1.5  0.055912
 0.2  1.5  2.0  0.052473
 0.2  2.0  0.5  0.021507
 0.2  2.0  1.0  0.020000
 0.2  2.0  1.5  0.018612
 0.2  2.0  2.0  0.017333
 0.4  0.5  0.5  1.369304
 0.4  0.5  1.0  1.264777
 0.4  0.5  1.5  1.174483
 0.4  0.5  2.0  1.096122
 0.4  1.0  0.5  0.450807
 0.4  1.0  1.0  0.400000
 0.4  1.0  1.5  0.356828
 0.4  1.0  2.0  0.320000
 0.4  1.5  0.5  0.194821
 0.4  1.5  1.0  0.168655
 0.4  1.5  1.5  0.146685
 0.4  1.5  2.0  0.128178
 0.4  2.0  0.5  0.093979
 0.4  2.0  1.0  0.080000
 0.4  2.0  1.5  0.068370
 0.4  2.0  2.0  0.058667
 0.6  0.5  0.5  1.771990
 0.6  0.5  1.0  1.549029
 0.6  0.5  1.5  1.375810
 0.6  0.5  2.0  1.239190
 0.6  1.0  0.5  0.735089
 0.6  1.0  1.0  0.600000
 0.6  1.0  1.5  0.498012
```

TP2 can be used to create a 3D plot of these results.

Chapter 22. Differential Equations

We will consider four types of differential equations in two groups of two. There are ordinary and partial differential equations. There are also initial value and boundary value problems. The combinations make four groups. There are linear and nonlinear differential equations, but this distinction is of little concern with numerical techniques. There are also individual equations and systems of equations. This distinction is also of little concern here, as we will focus on the methods and provide examples of how to handle these cases. We will consider stiff and loose systems, but these terms could apply to any in the four categories.

Ordinary differential equations basically have one temporal or spatial variable, for example: y(x) or x(t). Partial differential equations have more than one independent variable, for example: y(x,t) or p(x,y). The analytical techniques for solving these are often quite different, so it's not surprising that the numerical techniques would also be different. We begin our discussion with a simple and familiar case: a spring and mass. We could just as easily begin with a pendulum, which we will consider later.

A Simple First Order Ordinary Differential Equation

For the purpose of testing various methods, we will consider a very simple 1st order ordinary differential equation:

$$\frac{dy}{dx} = \sin(x) - \frac{y}{x} \qquad (22.1)$$

The analytical solution to this equation is:

$$y = \frac{\sin(x)}{x} - \cos(x) \qquad (22.2)$$

Second Order ODE: Damped Harmonic Oscillator

A mass connected to a spring and also a dashpot (damping device) is described by the following differential equation, where m is the mass, c is the damping factor, k is the spring constant, x is the displacement, and t is time.

$$m\frac{d^2x}{dt^2} + c\frac{dx}{dt} + kx = 0 \qquad (22.3)$$

The analytical solution is well known and begins by defining two additional parameters:

$$2\pi f_0 = \omega_0 = \sqrt{\frac{k}{m}} \qquad (22.4)$$

$$\beta = \frac{c}{m} \qquad (22.5)$$

The parameter $\omega_0 = 2\pi f_0$ is the resonant frequency (radians/second) and β is the damping coefficient. As many references explain, the solution to Equation 22.3 depends on the value of D in Equation 22.6.

$$D = \beta^2 - 4\omega_0 \qquad (22.6)$$

If the determinant D<0 is called under-damped, D>0 is called over-damped, and D=0 is called critically-damped. The under-damped solution is:

$$x = e^{-\frac{\beta t}{2}}\left[A\cos(\gamma t) + B\sin(\gamma t)\right] \qquad (22.7)$$

The coefficients A and B are determined by the initial conditions. The parameter γ is given by Equation 22.8:

$$\gamma = \frac{\sqrt{|D|}}{2} \qquad (22.8)$$

The over-damped solution is:

$$x = Ae^{\gamma_1 t} + Be^{\gamma_2 t} \qquad (22.9)$$

In this case, the paired parameter γ_{12} is given by Equation 22.10:

$$\gamma = \frac{-\beta \pm \sqrt{D}}{2} \qquad (22.10)$$

The critically-damped solution is given by:

$$x = (A + Bt)e^{-\omega_0 t} \qquad (22.11)$$

Three solutions ($\omega_0=1$, $\beta=0.5, 2, 3$) are shown in the next figure:

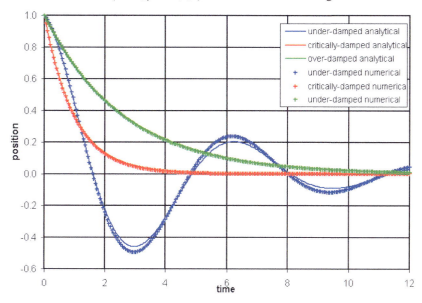

The analytical solutions are the solid lines and the first order (Euler's method) numerical solutions are the points. Agreement is quite good in this case for a time step $\Delta t=0.05$ seconds. The curves, calculations, and graph are provided in spreadsheet damped_harmonic_oscillator.xls in the on-line archive. We will use this system and analytical solutions to evaluate various numerical methods in the next chapter.

Third Order ODE

We will now consider the following simple third-order ordinary differential equation:

$$\frac{d^3 y}{dt^3} = -27 y \qquad (22.12)$$

The initial conditions are:

$$\begin{aligned} y(0) &= 4 \\ y'(0) &= -3 \\ y''(0) &= 9 \end{aligned} \qquad (22.13)$$

The solution is:

$$y = 2e^{-3t} + 2e^{\frac{3t}{2}} \cos\left(\frac{3t\sqrt{3}}{2}\right) \qquad (22.14)$$

Equation 22.14 plus a simple numerical solution is shown in the following figure:

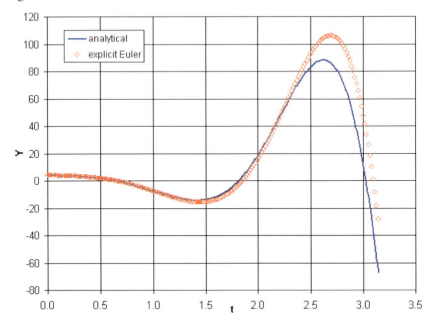

The interval above is 0 to π and the step size is $\pi/200$. Even with this small step size, the numerical solution doesn't match well. The higher order methods presented in Chapter 23 will do much better.

Chapter 23. Explicit Runge-Kutta Methods

Runge-Kutta is a type of marching method in that we start with some initial values and then step along through time (or space). We will first consider explicit methods. The seminal reference on the Runge-Kutta and related methods was published by Butcher.[29] There are countless articles on the Web dealing with Runge-Kutta. In marching methods, we consider differential equations of the following form:

$$\frac{dy}{dx} = f(x, y(x)) \qquad (23.1)$$

As we shall see, higher order differentials are easily handled by extending this formula. The initial position is represented by x and the time step, Δx, is represented by h. The symbol k is used to represent some particular value of $f(x,y(x))$. The simplest procedure is known as Euler's explicit method, which is implemented:

$$k_1 = f(x, y(x))$$
$$y(x+h) = y(x) + hk_1 \qquad (23.2)$$

This is exactly the same as:

$$y_{x+\Delta x} = y_x + \Delta x \left(\frac{dy}{dx}\right)_x \qquad (23.3)$$

Euler's explicit method is sometimes called 1st order Runge-Kutta. In general, these and similar methods can be expressed by the following formula, where n is the number of steps, which is not necessarily the same as the order:

$$k_1 = f(x, y)$$
$$k_2 = f(x + c_2 h, y + h(a_{21}k_1)) \qquad (23.4a)$$
$$k_3 = f(x + c_3 h, y + h(a_{31}k_1 + a_{32}k_2))$$
$$\ldots$$
$$k_i = f\left(x + c_i h, y + h\sum_{j=1}^{i-1} a_{ij}k_j\right) \qquad (23.4b)$$
$$y = y + h\sum_{i=1}^{n} b_i k$$

---

[29] Butcher, J. C., The Numerical Analysis of Ordinary Differential Equations: Runge-Kutta and General Linear Methods, John Wiley & Sons Ltd., New York, 1987.

Butcher Tableaus

Butcher expressed the preceding set of equations in tabular form, called a tableau, having the following form:

c_1	a_{11}	a_{12}	a_{13}	...	a_{1n}
c_2	a_{21}	a_{22}	a_{23}	...	a_{2n}
...	...	...	...	...	...
c_n	a_{n1}	a_{n2}	a_{n3}	...	a_{nn}
	b_1	b_2	b_3	...	b_n

The Butcher tableau for Euler's explicit method (Equation 23.2) is:

0	0
	1

We will present all of these methods in this way and then implement them in a code that can handle any formula in this form. There are three common variants of 2nd order Runge-Kutta. The first variant is:

0	0	0
1/2	1/2	0
	0	1

The second variant is called Huen's method:

0	0	0
1	1	0
	1/2	1/2

The third variant is called Ralston's method:

0	0	0
2/3	2/3	0
	1/4	3/4

There are also two common variants of 3rd order Runge-Kutta. The first is:

0	0	0	0
1/2	1/2	0	0
1	-1	2	0
	1/6	2/3	1/6

The second variant of 3rd order Runge-Kutta is:

0	0	0	0
1/3	1/3	0	0
2/3	0	2/3	0
	1/4	0	3/4

There are also two common variants of 4th order Runge-Kutta. The first is:

0	0	0	0	0
1/2	1/2	0	0	0
1/2	0	1/2	0	0
1	0	0	1	0
	1/6	1/3	1/3	1/6

This formula is reminiscent of Simpson's method for numerical integration. The second variant is:

0	0	0	0	0
1/3	1/3	0	0	0
2/3	-1/3	1	0	0
1	1	-1	1	0
	1/8	3/8	3/8	1/8

This formula is reminiscent of Simpson's 3/8ths rule.

Rabiei and Ismail[30] present a 5th order method which has been modified for inclusion here:[31]

0	0	0	0	0	0
0.25	0.25	0	0	0	0
0.25	-0.0086	0.2586	0	0	0
0.5	0.3868	-0.5312	0.6444	0	0
0.75	0.2067	-0.9002	0.8918	0.5517	0
	1.0222	0.0403	-0.107	-0.5999	0.6444

This is the only five-step 5th order method we will consider. Notice that the sum of the absolute value of the coefficients is greater than unity (e.g., 5111>5000). This is a sign that something is wrong. As with Newton-Cotes rules for numerical integration, the low-order methods are stable, but the higher-order methods are not. The stability deteriorates as the order increases.

Rabiei and Ismail also provide the following six-step 5th order method from Butcher:

0	0	0	0	0	0	0
1/4	1/4	0	0	0	0	0
1/4	1/8	1/8	0	0	0	0
1/2	0	-1/2	1	0	0	0
3/4	3/16	0	0	9/160	0	0
1	-3/7	2/7	12/7	-12/7	8/7	0
	7/90	0	32/90	12/90	32/90	7/90

Notice that orders one through four have the same number of steps as the order. When we come to order five, there are six steps (except for the first variant, Equation 23.14). This pattern continues to even higher orders. This is also reminiscent of Newton-Cotes rules (such as Simpson's). All is well with the lower orders, but higher orders seem to fall apart. The same thing happens with Runge-Kutta and for much the same reason. More details on fifth order

---

[30] Rabiei, F. and Ismail, F., "Fifth Order Improved Runge-Kutta Methods for Solving Ordinary Differential Equations," *Recent Researchers in Applied Informatics and Remote Sensing*, pp. 129-133, ISBN: 978-1-61804-039-8.
[31] There are several errors in the original publication. The sum of each row must meet the criteria: $c_i = \Sigma a_{ij}$ and $\Sigma b_i = 1$, which requires modifying several of the coefficients, including: changing a_{31} from 0.0082 to 0.0086, a_{41} from 0.3860 to 0.3868, a_{42} from +0.5312 to 0.5312, a_{51} from 0.2060 to 0.2067, and b_4 from -0.1 to -0.5999.

methods are provided by Luther and Konen.[32] Luther and Konen provide two more variants of 5th order Runge-Kutta. The first is:

0	0	0	0	0	0	0
1/3	1/3	0	0	0	0	0
2/5	4/25	6/25	0	0	0	0
1	1/4	-12/4	15/4	0	0	0
2/3	6/81	90/81	-50/81	8/81	0	0
4/5	6/75	36/75	10/75	8/75	0	0
	23/192	125/192	0	0	-81/192	125/192

The second 5th order variant provided by Luther and Konen is:

0	0	0	0	0	0	0
4/11	4/11	0	0	0	0	0
2/5	9/50	11/50	0	0	0	0
1	0	-11/4	15/4	0	0	0
$\frac{6-\sqrt{6}}{10}$	$\frac{81+9\sqrt{6}}{600}$	0	$\frac{255-55\sqrt{6}}{600}$	$\frac{24-14\sqrt{6}}{600}$	0	0
$\frac{6+\sqrt{6}}{10}$	$\frac{81-9\sqrt{6}}{600}$	0	$\frac{255+55\sqrt{6}}{600}$	$\frac{24+14\sqrt{6}}{600}$	0	0
	4/36	0	0	0	$\frac{16+\sqrt{6}}{36}$	$\frac{16-\sqrt{6}}{36}$

---

[32] Luther, H. A. and Konen, H. P., "Some Fifth-Order Classical Runge-Kutta Formulas," SIAM Review, Vol. 7, No. 4, 1965.

We will consider four variants of 6th order Runge-Kutta provided by Sarafyan.[33] The first is:

0	0	0	0	0	0	0	0	0
1/9	1/9	0	0	0	0	0	0	0
1/6	1/24	3/24	0	0	0	0	0	0
1/3	1/6	-3/6	4/6	0	0	0	0	0
1/2	1/8	0	0	3/8	0	0	0	0
2/3	-4/3	-21/3	46/3	-29/3	10/3	0	0	0
5/6	-8/72	99/72	-84/72	0	44/72	9/72	0	0
1	107/82	-243/82	0	354/82	-172/82	-36/82	72/82	0
	41/840	0	216/840	27/840	272/840	27/840	216/840	41/840

The second variant of 6th order Runge-Kutta is:

0	0	0	0	0	0	0	0	0
1/9	1/9	0	0	0	0	0	0	0
1/6	1/24	3/24	0	0	0	0	0	0
1/3	1/6	-3/6	4/6	0	0	0	0	0
1/2	1/8	0	0	3/8	0	0	0	0
2/3	6/3	-21/3	16/3	1/3	0	0	0	0
5/6	-68/72	99/72	96/72	-180/72	104/72	9/72	0	0
1	287/82	-243/82	-540/82	894/82	-352/82	-36/82	72/82	0
	41/840	0	216/840	27/840	272/840	27/840	216/840	41/840

---

[33] Sarafyan, D., "Improved Sixth-Order Runge-Kutta Formulas and Approximate Continuous Solution of Ordinary Differential Equations," *Journal of Mathematical Analysis and Applications*, Vol. 40, pp. 435-445, 1972.

The third variant of 6th order Runge-Kutta is:

0	0	0	0	0	0	0	0	0
1/9	1/9	0	0	0	0	0	0	0
1/6	1/24	3/24	0	0	0	0	0	0
1/3	1/6	-3/6	4/6	0	0	0	0	0
1/2	1/8	0	0	3/8	0	0	0	0
2/3	17/9	-63/9	51/9	0	1/9	0	0	0
5/6	-22/24	33/24	30/24	-58/24	34/24	3/24	0	0
1	281/82	-243/82	-522/82	876/82	-346/82	-36/82	72/82	0
	41/840	0	216/840	27/840	272/840	27/840	216/840	41/840

This last 6th order variant was developed by Huta.[34]

0	0	0	0	0	0	0	0	0
1/9	1/9	0	0	0	0	0	0	0
1/6	1/24	3/24	0	0	0	0	0	0
1/3	1/6	-3/6	4/6	0	0	0	0	0
1/2	-5/8	27/8	-24/8	6/8	0	0	0	0
2/3	221/9	-981/9	867/9	-102/9	1/9	0	0	0
5/6	-183/48	678/48	-472/48	-66/48	80/48	3/48	0	0
1	716/82	-2079/82	1002/82	834/82	-454/82	-9/82	72/82	0
	41/840	0	216/840	27/840	272/840	27/840	216/840	41/840

We will now evaluate these 16 methods (3s 2.6 through 23.20) against the three analytical solutions presented in Chapter 22. The code to implement these tableaus is listed in Appendix G. The tableaus along with the code to solve the three examples from Chapter 22 can be found in on-line archive in file RKcomparison.c.

---

[34] Huta, A., "Contribution a la Formule de Sixième Ordre dans la Methodè de Runge-Kutta-Nyström (Contribution to the Sixth-Order Formula in the Runge-Kutta-Nyström Method)," Acta Fat. Rerum Natur. Univ. Cominian. (University Faculty Math Lectures), Vol. 2, pp. 21-23, 1957.

First Order ODE Test

Solving Equation 22.1 from x=0 to x=3 with a step size h=0.1 for each of the 16 methods yields the following table:

equation	order	result	error
1.2	analytical	1.03703	N/A
2.6	1st	1.06468	2.7%
2.7	2nd-1	1.03737	0.0%
2.8	2nd-2	1.03625	-0.1%
2.9	2nd-3	1.03700	0.0%
2.10	3rd-1	1.03703	0.0%
2.11	3rd-2	1.03699	0.0%
2.12	4th-1	1.03703	0.0%
2.13	4th-2	1.03703	0.0%
2.14	5th-1	1.05505	1.7%
2.15	5th-2	1.03703	0.0%
2.16	5th-3	1.03942	0.2%
2.17	5th-4	1.03704	0.0%
2.18	6th-1	1.03703	0.0%
2.19	6th-2	1.03702	0.0%
2.20	6th-3	1.03703	0.0%
2.21	6th-5	1.03703	0.0%

It is not surprising that the 1st order has an error of 2.7%, nor that one of the 2nd order methods (Equation 23.8) has an error of -0.1%. It is surprising that the first 5th order method (Equation 23.14) has an error of 1.7%. This is after fixing several of the terms. Clearly, there is still something wrong with it. We could try other values, but there's no point with so many other methods to choose from. Somewhat surprising is the 0.2% error of the third 5th order method (Equation 23.16). All of the other methods are accurate within 5 to 6 significant figures.

Solving High Order Differential Equations

Differential equations of higher order (i.e., above one) are implemented as an array. The size of the array is equal to the order of the differential equation. The highest order differential goes in the highest index of the array. The next differential is equal to the high order solution. Equation 22.1 becomes:

```
dY[0]=sin(X)-Y[0]/X;
```

Second order ODE Equation 22.3 becomes:

```
dY[1]=-w0*w0*Y[0]-beta*Y[1];
dY[0]=Y[1];
```

Third order ODE Equation 22.12 becomes:

```
dY[2]=-27.*Y[0];
dY[1]=Y[2];
dY[0]=Y[1];
```

This same pattern is repeated to whatever level is required to specify the differential equation. You can also solve parallel equations having different orders. For example, a third plus a second order differential equation would be entered as:
```
dY[4]=f(x,...);
dY[3]=Y[4];
dY[2]=Y[3];
dY[1]=g(x,...);
dY[0]=Y[1];
```

The first two elements of the arrays (Y[0], Y[1], dY[0], dY[1]) contain the second equation (2nd order ODE) and the next three elements (Y[2], Y[3], Y[4], dY[2], dY[3], dY[4]) contain the second (3rd order ODE).

Second Order ODE Test

The next test is solve Equation 22.3 with the three values of β (0.5,2,3) illustrated in the figure at the end of Chapter 22 using each of the 16 methods. The RungeKutta() function is called 240 times for each value of β. The microsecond timer is called before and after so as to compare the computational effort for each method. Results of this test are shown in the next table:

		h=0.05 (240 steps)			
equation	order	error	μsec	Δμsec	1/Δμsec/err
1.7,9,11	analytical	N/A	90.8	N/A	N/A
2.6	RK1	5.28118000000000	119.6	28.8	0.01
2.7	RK2a	0.08128200000000	146.7	55.9	0.22
2.8	RK2b	0.08128200000000	145.8	55.0	0.22
2.9	RK2c	0.08128200000000	145.8	55.0	0.22
2.10	RK3a	0.00098296500000	185.8	95.0	10.71
2.11	RK3b	0.00098296500000	177.4	86.6	11.75
2.12	RK4a	0.00000958646000	234.7	143.9	725.04
2.13	RK4b	0.00000958646000	234.1	143.3	727.87
2.14	RK5a	3.47677000000000	403.7	312.9	0.0009
2.15	RK5b	0.00000001055510	319.0	228.2	415,091
2.16	RK5c	0.43218100000000	318.2	227.4	0.01
2.17	RK5d	0.00000008270320	316.2	225.4	53,633
2.18	RK6a	0.00000000026026	433.0	342.2	11,227,488
2.19	RK6b	0.01304280000000	434.1	343.3	0.22
2.20	RK6c	0.00000000008336	445.9	355.1	33,784,055
2.21	RK6d	0.00000000007166	433.3	342.5	40,744,544

We subtract the analytical time (90.8 microseconds) from each of the others in order to get the time devoted to the R-K algorithm alone (e.g., 119.6-90.8=28.8 for RK1). We divide 1 by time and error to get a sort of efficiency (shorter time is better and so is less error). All of the 1st, 2nd, and 3rd order methods come out poorly in this test by orders of magnitude. The RK5a, RK5c, and RK6b also come out very poorly, which is why we don't ever want to use any of these methods. They're interesting for theory and history, but useless in practice. A section of the code used to produce this table listed below:

```
double compare1(double b,double*BT,int steps)
  {
  int i;
  double dt=0.05,s,t,x[2];
  beta=b;
  s=t=0.;
  x[0]=1.;
  x[1]=begin();
  for(i=1;i<=240;i++)
    {
    RungeKutta(BT,steps,dxdt,&t,dt,x,2);
    s+=fabs(x[0]-oscillator(t));
    }
  return(s);
  }
void compare3(char*name,double*BT,int steps)
  {
  int j;
  double s;
  __int64 rate,t1,t2;
  QueryPerformanceFrequency(&rate);
  QueryPerformanceCounter(&t1);
  for(s=j=0;j<3;j++)
    s+=compare1(betaj[j],BT,steps);
  QueryPerformanceCounter(&t2);
  printf("%s\t%lG\t%lG\n",name,s,1E6*(t2-
    t1)/((double)rate));
  }
void test2b()
  {
  compare3("analytical",NULL,0);
  compare3("RK1"  ,RK1 ,1);
  compare3("RK2a",RK2a,2);
  compare3("RK2b",RK2b,2);
  compare3("RK2c",RK2c,2);
etc...
```

We will now double the time step (h=0.1) and cut the number of steps in half to arrive at the same final result, producing the following table:

		h=0.1 (120 steps)				error
equation	order	error	μsec	Δμsec	1/Δμsec/err	h=0.05/0.10
1.7,9,11	analytical	N/A	97.2	N/A	N/A	N/A
2.6	RK1	5.65166000000000	125.7	28.5	0.01	0.934
2.7	RK2a	0.16463500000000	150.3	53.1	0.11	0.494
2.8	RK2b	0.16463500000000	299.5	202.3	0.03	0.494
2.9	RK2c	0.16463500000000	149.7	52.5	0.12	0.494
2.10	RK3a	0.00398250000000	184.1	86.9	2.89	0.247
2.11	RK3b	0.00398250000000	180.7	83.5	3.01	0.247
2.12	RK4a	0.00007793300000	230.8	133.5	96.09	0.123
2.13	RK4b	0.00007793300000	250.6	153.4	83.66	0.123
2.14	RK5a	3.66533000000000	350.9	253.7	0.0011	0.949
2.15	RK5b	0.00000017534000	570.5	473.2	12,051	0.060
2.16	RK5c	0.43468700000000	334.4	237.2	0.01	0.994
2.17	RK5d	0.00000134235000	360.9	263.7	2,825	0.062
2.18	RK6a	0.00000000853754	454.2	357.0	328,068	0.030
2.19	RK6b	0.02619550000000	525.5	428.3	0.09	0.498
2.20	RK6c	0.00000000275836	460.4	363.2	998,236	0.030
2.21	RK6d	0.00000000237685	442.0	344.7	1,220,423	0.030

The far right column is the error at h=0.05 compared to the error at h=0.1. One might hope that this result would be close to 0.5 (i.e., half the time step yielding half the error). It is for RK2a, RK2b, RK2c, and RK6b, but not so for every method. This brings up an important issue and a new expectation. The best methods are: RK4a, RK4b, RK5b, RK5d, RK6a, RK6c, and RK6d (the same ones that came out on top in the previous test). This is why you hardly ever see anyone implementing any of the other methods. They're only discussed from a theoretical perspective. Using this same table we can further narrow the selection eliminating RK6a in favor of RK6c and RK6d.

Third Order ODE Test

The next test is solve Equation 22.12 over the range 0 to π with a step size of $\pi/50$. The results are shown in the following table:

equation	order	error	μsec	Δμsec	1/Δμsec/err
			h=π/50 (50 steps)		
1.14	analytical	N/A	34.9	N/A	N/A
2.6	RK1	1439.7	36.9	2.0	0.00035
2.7	RK2a	52.8459	36.3	2.0	0.0095
2.8	RK2b	52.8459	34.4	2.0	0.0095
2.9	RK2c	52.8459	48.6	13.7	0.0014
2.10	RK3a	2.98706	43.3	8.4	0.040
2.11	RK3b	2.98706	40.5	5.6	0.060
2.12	RK4a	0.140357	61.2	26.3	0.27
2.13	RK4b	0.140357	50.0	15.1	0.47
2.14	RK5a	892.744	99.2	64.3	0.000017
2.15	RK5b	0.000371488	70.7	35.8	75
2.16	RK5c	122.484	69.3	34.4	0.00024
2.17	RK5d	0.00287056	66.8	31.8	11
2.18	RK6a	0.0000394482	98.9	64.0	396
2.19	RK6b	8.62185	89.7	54.8	0.0021
2.20	RK6c	0.0000118713	95.0	60.1	1,402
2.21	RK6d	0.00000997064	90.2	55.3	1,813

A similar pattern as before can be seen here. The same methods prove superior to the rest (RK4a, RK4b, RK5b, RK5d, RK6a, RK6c, and RK6b). In the next chapter we will consider some more interesting applications of these methods.

Chapter 24. Marching Method Applications

In the last chapter we compared the various Runge-Kutta methods and found 7 out of 16 to be superior to the rest (RK4a, RK4b, RK5b, RK5d, RK6a, RK6c, and RK6b). The last two are by far the most efficient. We will now consider some useful examples.

Initial Value Problem: Young-Laplace Capillary Equation

This partial differential equation describes the forces acting on a membrane or the interface between a liquid and vapor. It relates the pressure differential to the curvature:

$$\Delta p = \frac{\sigma}{\dfrac{1}{R_1} + \dfrac{1}{R_2}} \tag{24.1}$$

In this equation R_1 and R_2 are the two radii of curvature and σ is the surface tension. The solution we seek is the shape of a sessile droplet.[35] The shape and variables are illustrated in the following figure:

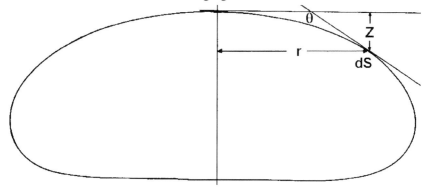

In this figure, r is the local radius of the droplet, z is the vertical distance, θ us the angle of the surface, and ds is the distance along the surface. The initial conditions are $r=z=\theta=0$. Equation 24.1 can be recast in terms of the variables in this figure to become:

$$\frac{dr}{d\theta} = \frac{\cos\theta}{\dfrac{2}{b} + z - \dfrac{\sin\theta}{r}} \tag{24.2}$$

---

[35] The web is filled with botanical uses of the word *sessile*, but these have nothing to do with the definition used here. This word comes from the latin *seated*, which is how it's used here.

$$\frac{dz}{d\theta} = \tan\theta \frac{dr}{d\theta} \qquad (24.3)$$

$$\frac{dV}{d\theta} = \pi r^2 \frac{dz}{d\theta} \qquad (24.4)$$

$$b = \left.\frac{dz^2}{d\theta^2}\right|_{\theta=0} \qquad (24.5)$$

The parameter b is a constant equal to the second derivative of the vertical distance at the top/center. Different values of b produce different size droplets. The parameter V is the droplet volume. The vertically projected area is equal to π times the maximum radius squared. We will solve this differential equation using 6th order Runge-Kutta (method RK6d). The equations are entered as follows:

```
#define  R   Y[0]
#define  Z   Y[1]
#define  V   Y[2]
#define  dR  dY[0]
#define  dZ  dY[1]
#define  dV  dY[2]
void drop(double theta,double*Y,double*dY)
  {
  double sr,dS;
  theta=min(theta,M_PI);
  dS=1./b+Z;
  if(fabs(R)>DBL_EPSILON)
    {
    sr=sin(theta)/R;
    if(sr>0.&&sr<1./b)
      dS=2./b+Z-sr;
    }
  dR=cos(theta)/dS;
  dZ=sin(theta)/dS;
  dV=M_PI*R*R*dZ;
  }
```

The source code can be found in the on-line archive in file sessile.c. This has already been compiled and will run on any version of Windows®. There's also a little batch file _compile_sessile.bat to recompile it if desired. Note the 6 #define statements make it easier to keep track of the variables and to express the formulas in the original (non-coded) terms. The results can be found in file sessile.out and sessile.xls. The resulting shapes are shown in this next figure:

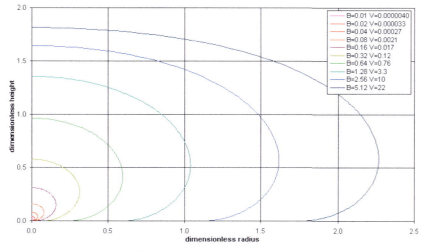

Initial Value Problem: Car Acceleration

No discussion of differential equations would be complete without the first such problem I ever solved numerically: car acceleration. We'll need a torque or horsepower curve:

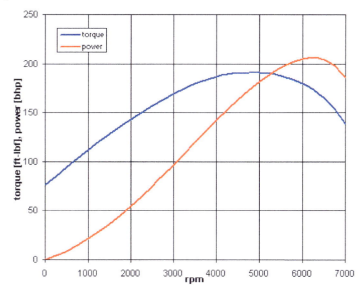

Note that *horsepower=torque*2π*rpm/550/60* so that if you have one curve, the other is easily calculated. We will also need a mass (2861 pounds) and gear ratios:

gear	ratio
1	3.592
2	2.057
3	1.361
4	1.000
5	0.821

We also need to know the axle ratio and tire diameter. The rpm at 60 mph in overdrive will also suffice (2200). We also need to decide the stall and shift speeds (1250 and 6250 rpm). The rest is calculus! Abbreviated results are listed in the following table:

time	rot	torque	gear	accel	veloc	speed	dist
sec	rpm	ft-lbf	shift	ft/sec^2	ft/sec	mph	ft
0.00	1250	120.3	1	32.17	0.0	0.0	0
1.00	2305	151.9	1	19.64	21.1	14.4	12
2.00	4796	191.2	1	24.71	43.9	29.9	44
3.00	3968	186.1	2	13.78	63.3	43.2	99
4.00	4846	191.2	2	14.15	77.4	52.8	169
5.00	5722	184.6	2	13.66	91.4	62.3	253
6.00	4278	189.1	3	9.26	103.2	70.4	351
7.00	4664	191.0	3	9.35	112.5	76.7	459
8.00	5052	190.7	3	9.34	121.9	83.1	576
9.00	5437	188.2	3	9.22	131.2	89.4	703
10.00	5814	183.1	3	8.97	140.3	95.7	839
11.00	6178	175.2	3	8.58	149.1	101.6	983
12.00	4760	191.1	4	6.88	156.3	106.6	1136
13.00	4969	191.0	4	6.87	163.2	111.3	1296
13.15	5001	190.9	4	6.87	164.2	112.0	1320

All the details can be found in the on-line archive in spreadsheet car_simulation.xls. This problem is solved in the spreadsheet using the explicit Euler method with a small time step (0.01 sec). You can change values in the spreadsheet and the simulation will update automatically. This problem can also be solved using Runge-Kutta methods. The source code can be found in car_simulation.c. Due to the gear shifting, the implementation is a little more complicated than the previous problems, but still rather simple:

```
double torque(double rpm)
  {
  return ((((((-2.39436962861068E-
    24*rpm+4.94254828375489E-20)*rpm
    -3.88505184269642E-16)*rpm+1.39955406645336E-12)*rpm
    -2.54306117072706E-9)*rpm-1.88969627673033E-19)*rpm
    +0.037365290797588)*rpm+76.1035566976854;
```

```
    }

double horsepower(double rpm)
  {
  return torque(rpm)*rpm*2.*M_PI/550./60.;
  }

double g=32.174;
double weight=2851.;
double rpmat60=2200.;
double rpm_min=1250.;
double rpm_max=6250.;
double ratios[]={3.592,2.057,1.361,1.,0.821};
#define ngear (sizeof(ratios)/sizeof(ratios[0]))

double rpms(int gear,double fps)
  {
  double rpm;
  rpm=rpmat60*(fps/88.)*ratios[gear]/ratios[ngear-1];
  if(rpm>=rpm_min)
    return rpm;
  return rpm_min;
  }

int gears(double fps)
  {
  int gear=0;
  while(fps>0.&&gear<ngear-1)
    {
    if(rpms(gear,fps)<rpm_max)
      break;
    gear++;
    }
  return(gear);
  }

int gear;
double rpm,torq;

void car(double t,double*y,double*dy)
  {
  gear=gears(y[1]);
  rpm=rpms(gear,y[1]);
  torq=torque(rpm);
  dy[1]=min(g,torq*rpm*2.*M_PI/60./max(y[1],
    DBL_EPSILON)/(weight/g));
  dy[0]=y[1];
  }
```

Note that you must handle the case of $V=0$ so as to not divide by zero. The results are shown in the following figure.

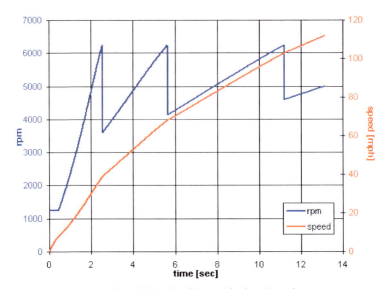

Boundary Value Problem: Blasius Equation

This problem arises from the partial differential equation describing fluid flow near a fixed surface. This near-wall region is call the boundary layer. A detailed discussion of this topic can be found in any fluid mechanics text. A recent reference readily available on-line is the paper by Jaman, Molla, and Sultana.[36] A transformation of variables is applied to the governing equation that results in the following problem:

$$2y'' + yy' = 0$$
$$y(0) = y'(0) = 0 \qquad (24.6)$$
$$y'(\infty) = 1$$

As usual, y' indicated dy/dx and y'' indicates d^2y/dx^2. This is not an initial value problem, rather it is a boundary value problem. Still, it can be solved as an initial value problem by guessing the value of $y'(0)$ that results in $y'(\infty)=1$. Guessing an initial value to arrive at an eventual result is called *shooting*. There are may problems that can be solved in this way. For more on this the reader is directed to the paper by Summiya.[37]

Equation 24.6 is easily implemented as follows:

---

[36] Jaman, M. K., Molla, M. R., and Sultana, S., Numerical Approximations of Blasius Boundary Layer Equation, Dhaka University Journal of Science, Vol. 59, No. 1, pp. 87-90, 2011.

[37] Summiya, P., Numerical Solution of Non Linear Differential Equation by Using Shooting Techniques, International Journal of Mathematics and Its Applications, Vol. 4, No. 1-A, pp. 93-100, 2016.

```
dY[2]=-Y[0]*Y[2]/2.;
dY[1]=Y[2];
dY[0]=Y[1];
```

We will use a 4th and a 6th order method to solve this problem (Equations 23.12/RK4a and 23.21/RK6d). The code can be found in the on-line archive in file blasius.c. This has already been compiled and will run on any version of Windows®. A small batch file, _compile_blasius.bat, is also provided should you want to recompile it. The results can be found in blasius.xls and are listed below in abbreviated form. The Excel® spreadsheet contains an explicit and implict Euler solution as well.

```
                             BLASIUS
          4th Order Runge-Kutta      6th Order Runge-Kutta
   X       Y      dY/dX   dY²/d²X      Y      dY/dX   dY²/d²X
  0.0   0.00000  0.00000  0.33206    0.00000  0.00000  0.33206
  0.4   0.02656  0.13276  0.33147    0.02656  0.13276  0.33147
  0.8   0.10611  0.26471  0.32739    0.10611  0.26471  0.32739
  1.2   0.23795  0.39378  0.31659    0.23795  0.39378  0.31659
  1.6   0.42032  0.51676  0.29666    0.42032  0.51676  0.29666
  2.0   0.65002  0.62977  0.26675    0.65002  0.62977  0.26675
  4.0   2.30575  0.95552  0.06423    2.30575  0.95552  0.06423
  6.0   4.27962  0.99897  0.00240    4.27962  0.99897  0.00240
  8.0   6.27921  1.00000  0.00001    6.27921  1.00000  0.00001
 10.0   8.27921  1.00000  0.00000    8.27921  1.00000  0.00000
 12.0  10.27921  1.00000  0.00000   10.27921  1.00000  0.00000
```

A More Complex Problem: Parallel Equations

The next problem we consider (a thermal plume discharged into a flowing river) is more complex because there are several equations to solve in parallel, including: conservation of mass, momentum, and energy, as well as width and position. There is also the conservation of salt if this is a brine plume or the receiving water is salty. The model is described in detail in my TWRA paper.[38]

There is considerably more input and output associated with this problem. There is the composition of the plume and the ambient: temperature, salinity, and velocity. The plume may also be discharged from a slot or a round jet and the equations must handle both cases. We also want to display the results graphically, so we will create all of the necessary files to accomplish this. The entire code, along with four sample input files (plume1.txt, plume2.txt, plume3.txt, and plume4.txt), can be found in the on-line archive in files plume2d.*. The variables are listed in the following table:

---

[38] Benton, D. J., "Development of a Two-Dimensional Plume Model for Positively and Negatively Buoyant Discharges into a Stratified Flowing Ambient," *Tennessee Water Resources Symposium*, 1989.

variable	units	description
α	-	entrainment coefficient
b	ft	width/diameter of the plume
depth	ft	depth
difd	ft	diffuser diameter
difl	ft	diffuser length
dify	ft	diffuser elevation (up from bottom)
dil	-	mixing ratio (dilution)
Froude	-	densimetric Froude number
Qr	ft^3/sec	river flow
Qdis	ft^3/sec	discharge flow
ρ	lbm/ft^3	density
S	-	salinity
T	°F	temperature
ang	°	angle of inclination from the horizontal
u	ft/s	horizontal velocity of the plume
v	ft/s	vertical velocity of the plume
w	ft/s	velocity along centerline w=sqrt(u^2+v^2)
x	ft	horizontal distance from diffuser ports
y	ft	vertical distance from diffuser ports
z	ft	distance along the centerline

The governing equations in symbolic form are:

p(1)=x	horizontal coordinate of centerline
p(2)=y	vertical coordinate of centerline
p(3)=r*w*b^2*π/4	round jet mass flux
p(3)=r*w*b	slot jet mass flux/unit diffuser length
p(4)=p(3)*u	horizontal momentum flux
p(5)=p(3)*v	vertical momentum flux
p(6)=p(3)*t	thermal energy flux
p(7)=p(3)*s	salt flux

The implementation is more lengthy too:
```
double ftr(double y)
 {/* river temperature interpolation function */
  int i;
  if(y>=yr[0])
    return tr[0];
  for(i=1;i<ntr;i++)
    if(y<=yr[i-1]&&y>=yr[i])
```

```
      return tr[i-1]+(y-yr[i-1])*(tr[i]-tr[i-
    1])/(yr[i]-yr[i-1]);
  return tr[ntr-1];
  }
double fur(double y)
  {/* river velocity interpolation function */
  int i;
  if(nur<1)
    return urave;
  if(y>=yr[0])
    return ur[0];
  for(i=1;i<nur;i++)
    if(y<=yr[i-1]&&y>=yr[i])
      return ur[i-1]+(y-yr[i-1])*(ur[i]-ur[i-
    1])/(yr[i]-yr[i-1]);
  return ur[nur-1];
  }
double fsr(double y)
  {/* river salinity interpolation function */
  int i;
  if(nsr<1)
    return srave;
  if(y>=yr[0])
    return sr[0];
  for(i=1;i<nsr;i++)
    if(y<=yr[i-1]&&y>=yr[i])
      return sr[i-1]+(y-yr[i-1])*(sr[i]-sr[i-
    1])/(yr[i]-yr[i-1]);
  return sr[nsr-1];
  }
void plume(double z,double*p,double*dp)
  {/* the differentials */
  double alpha,b,froude,ramb,rplm,samb,splm,tamb,tplm,
    uamb,uave,vave,vplm,y;
  /* check for unstable plume */
  if(p[2]<=0.)
    {
    dp[2]=0.;
    dp[3]=0.;
    dp[4]=0.;
    dp[5]=0.;
    dp[6]=0.;
    return;
    }
  /* calculate plume velocities */
  y=p[1];
  if(round)
    {
    uave=p[3]/p[2]/1.33;
    vave=p[4]/p[2]/1.33;
```

```
      }
    else
      {
      uave=p[3]/p[2]/1.43;
      vave=p[4]/p[2]/1.43;
      }
  vplm=sqrt(uave*uave+vave*vave);
  /* calculate plume trajectory angle */
  dp[0]=uave/vplm;
  dp[1]=vave/vplm;
  /* calculate plume temperature,density,and width */
  tplm=p[5]/p[2];
  splm=p[6]/p[2];
  rplm=rho(tplm,splm);
  if(round)
    b=sqrt(4.*p[2]/rplm/vplm/M_PI);
  else
    b=p[2]/rplm/vplm;
  tamb=ftr(y);
  uamb=fur(y);
  samb=fsr(y);
  ramb=rho(tamb,samb);
  if(round)
    {
    froude=vplm/sqrt(fmax(0.000001,fabs(g*(ramb-
    rplm)*b/2./ramb)));
    if(acons>0.)
      alpha=acons;
    else
      alpha=0.0535*exp(1.43/sq(fmax(3.2,fmin(16.,
    froude))));
    dp[2]=M_PI*b*alpha*ramb*hypot(uamb-uave,vave);
    dp[3]=dp[2]*uamb;
    dp[4]=(ramb-rplm)*g*M_PI*b*b/4.;
    dp[5]=tamb*dp[2];
    dp[6]=samb*dp[2];
    }
  else
    {
    froude=vplm/sqrt(fmax(0.000001,fabs(g*(ramb-
    rplm)*b/ramb)));
    if(acons>0.)
      alpha=acons;
    else
      alpha=0.0520*exp(1.62/pow(fmax(2.4,fmin(20.,
    froude)),1.5));
    dp[2]=alpha*ramb*hypot(uamb-uave,vave);
    dp[3]=dp[2]*uamb;
    dp[4]=(ramb-rplm)*g*b;
    dp[5]=tamb*dp[2];
```

```
        dp[6]=samb*dp[2];
    }
}
```

The system of equations is solved using 4th order Runge-Kutta (RK4c). Typical input is listed below:

```
test case#1 specified river flow/uniform velocity
50 46000 60000  depth,river cross-section,river flow
118.8 0 2500 700 17 .37 0 33 0  Tdis,Sdis,Qdis,difl,
    difd,dify,slotwidth,angle,alpha
10 0 0   number of temperatures, number of velocities
46.7 71.9 0 0   (z,t,v,s) if(nv<>0) (read in velocities)
45.1 71.4 0 0   if(nv=nt)then ignore velocities
43.4 70.4 0 0
42.6 69.4 0 0
39.3 68.9 0 0
36.0 68.5 0 0
26.2 67.9 0 0
16.3 67.5 0 0
 6.5 67.2 0 0
 0.0 67.0 0 0
```

Results for the first example (a rising plume) are shown in this figure:

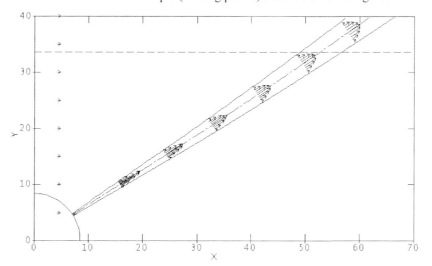

Results for the fourth example (a sinking plume) are shown in this next figure:

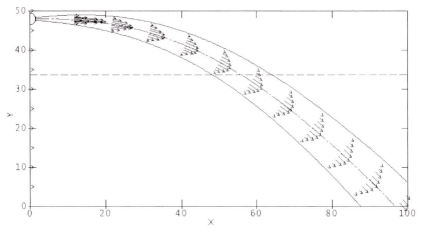

Crossflow Cooling Tower

The next problem we consider is combined heat and mass transfer within the packing inside a crossflow evaporative cooling tower. This next figure illustrates a crossflow cooling tower:

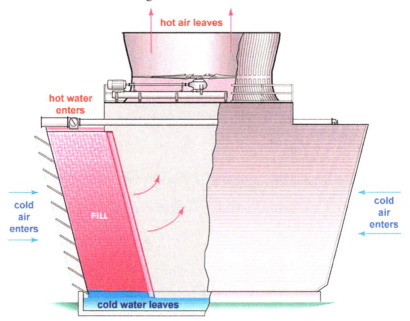

The fill material varies, but often looks something like this:

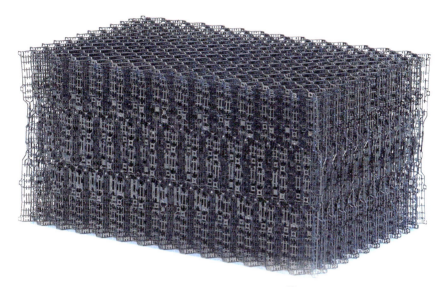

The governing equation goes back to Merkel.[39] In differential form the conservation of energy within a cell can be expressed by:

$$GKaY(h_W - h_A) - LCdT_W = 0 \qquad (24.7)$$

In this equation G is the mass flux of air (lb/ft²/hr or kg/m²/s), a is the interfacial area per unit volume (ft²/ft³ or m²/m³), Y is the height of the packing (ft or m), h_W and h_A are the enthalpies of saturated moist air at the temperature of the water and air, respectively (BTU/lb or kJ/kg), L is the mass flux of the water (lb/ft²/hr or kg/m²/s), C is the specific heat of the water (BTU/lb/°F or kJ/kg/°C), and T_W is the temperature of the water (°F or °C). The factor K in Equation 24.7 is the dimensionless mass transfer coefficient. Merkel was interested in counterflow cooling towers and used the 4-point Chebyshev method to solve the equation. Here, we will consider crossflow. The differential is quite simple:

```
void Cell(double X,double Y,double Ha,double*dHa,double
   Tw,double*dTw)
{
double Hw,Q;
Hw=fHtwb(user.baro,Tw);
Q=user.KaY*(Hw-Ha);
dHa[0]=Q*user.LG;
dTw[0]=-Q;
}
```

---

[39] Merkel, F. Verdunstungskulung, V.D.I. Forschungsarbeiteh (Society of German Engineers Technical Journal), No. 275, Berlin, 1925.

We start at the upper left hand corner of a rectangular block of cells. The hot water enters this cell from the top and the ambient air enters from the left. We can then use 4th order Runge-Kutta to integrate across and down, also very simple:

```
for(y=0;y<Ny;y++)
   {
   for(x=0;x<Nx;x++)
      {
      H=Ha[(Nx+1)*y+x];
      T=Tw[Nx*y+x];
      RungeKutta2D(Cell,X,1./Nx,Y,1./Ny,&H,&T);
      Ha[(Nx+1)*y+x+1]=H;
      Tw[Nx*(y+1)+x]=T;
      }
   }
```

The air and water temperatures (Ta and Tw) along with the air and water enthalpies (ha and hw) for a 5x5 grid are illustrated below:

		Ta				Tw				
						148	148	148	148	148
78	102	116	124	128	132	115	115	124	130	131
78	92	99	108	115	118	97	104	108	113	121
78	86	92	98	103	108	88	96	99	104	110
78	82	88	92	96	101	84	89	93	98	102
78	80	84	87	91	94	81	85	89	92	97
						11 approach				89
								Hw		
		Ha				260.1	260.1	260.1	260.1	260.1
41.6	74.7	107.4	131.5	149.2	165.8	104.7	105.8	132.7	157.3	162.5
41.6	59.3	70.8	86.9	103.7	114.1	66.8	79.0	87.5	100.9	122.9
41.6	50.5	58.7	68.0	77.6	88.2	53.6	64.4	69.2	79.0	93.4
41.6	46.0	52.8	58.6	64.5	73.4	48.0	54.5	59.9	68.0	74.5
41.6	44.0	48.0	51.8	57.5	62.2	45.2	49.3	54.6	59.1	66.2

This calculation is used to draw what are called *demand* curves. These tell to what value of K is required to the water to a certain temperature at a certain ratio of water to air flux. The packing (various sorts of fill material, most often plastic) provides some level of performance, which we call *supply*. The intersection of the supply and demand curves is where the cooling tower will operate. This is how cooling towers are designed.

We start with a value of entering air temperature (cool ambient, not necessarily uniform across the vertical face of the packing), entering water temperature (hot and uniform across the top of the fill), and a value of KaY/L. Then we step through the fill using Runge-Kutta to solve the differential equation, ultimately obtaining an average exiting (cooled) water temperature. We want curves of equal exiting water temperature, or more precisely, equal approach (exiting water temperature minus entering wet-bulb temperature). We adjust the value of KaY/L in order to match the desire approach and eventually draw a curve.

To accomplish this we use a bisection search. We pick a lower and upper limit on KaY/L and then iterate, each time cutting the interval in half until the result is within $1/2^{32}$ of the original span of KaY/L. This simple technique is very useful and works in many cases to solve nonlinear problems. Typical curves are:

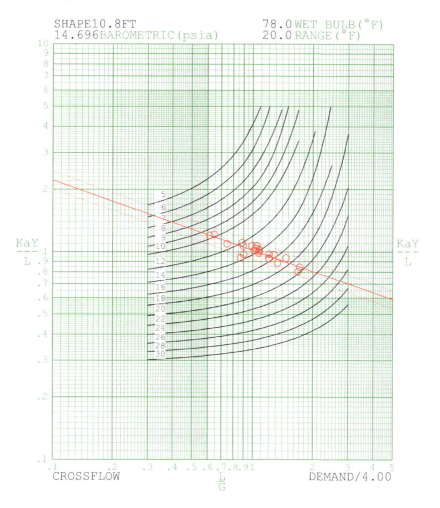

You will find a program (with source code) in the on-line archive in the folder examples\KaVL that is interactive, solves this problem for you, and creates graphs in English or SI units. The input dialog is:

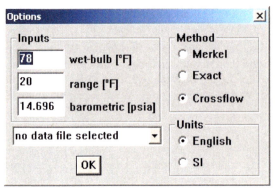

The resulting graph is pasted onto the clipboard. The code to produce the graph and copy it to the clipboard is also useful in other applications. Feel free to use it!

Chapter 25. Step Length Control

Considerable effort has been devoted to arriving at an error estimate for various Runge-Kutta methods so as to control the step length automatically. The standard approach is to calculate the result (i.e., the next step in the dependent variable) with two similar methods, one having a higher order than the other. We could simply use RK3a and RK4a or some other such combination; however, a desire to make this process more efficient has motivated theorists to find methods that use the same intermediate steps. For this reason these procedures are sometimes called *embedded* methods, although *parallel* methods might be more descriptive. We will consider three such methods. The first is called Huen-Euler and has the following Butcher tableau:

0	0	0
1	1	0
	1/2	1/2
	1	0

Instead of a single bottom row containing b_i, there are two such rows, one for each of the estimates. The second method is attributed to Fehlberg:

0	0	0	0
1/2	1/2	0	0
1	-1/256	255/256	0
	1/256	255/256	0
	1/512	255/256	1/512

The third is called Bogacki-Shampine:

0	0	0	0	0
1/2	1/2	0	0	0
3/4	0	3/4	0	0
1	2/9	1/3	4/9	0
	2/9	1/3	4/9	
	7/24	1/4	1/3	1/8

The following code implements each of these methods:

```
double HuenEuler(void dYdX(double,double*,
    double*),double*X,double dX,double*Y)
  {
  double k1,k2,W;
```

```
  dYdX(*X,Y,&k1);
  W=*Y+k1*dX;
  dYdX(*X+dX,&W,&k2);
  *Y=*Y+(k1+k2)*dX/2.;
  *X=*X+dX;
  return(*Y-W);
  }
double Fehlberg(void dYdX(double,double*,
    double*),double*X,double dX,double*Y)
  {
  double k1,k2,k3,W;
  dYdX(*X,Y,&k1);
  W=*Y+k1*dX/2.;
  dYdX(*X+dX/2.,&W,&k2);
  W=*Y+(k1+255.*k2)*dX/256.;
  dYdX(*X+dX,&W,&k3);
  *Y=*Y+(k1+510.*k2+k3)*dX/512.;
  *X=*X+dX;
  return(*Y-W);
  }
double BogackiShampine(void dYdX(double,double*,
    double*),double*X,double dX,double*Y)
  {
  double k1,k2,k3,k4,W;
  dYdX(*X,Y,&k1);
  W=*Y+k1*dX/2.;
  dYdX(*X+dX/2.,&W,&k2);
  W=*Y+3.*k2*dX/4.;
  dYdX(*X+3.*dX/4.,&W,&k3);
  W=*Y+(2.*k1+k2+4.*k3)*dX/9.;
  dYdX(*X+dX,&W,&k4);
  *Y=*Y+(7.*k1+6.*k2+8.*k3+3.*k4)*dX/24.;
  *X=*X+dX;
  return(*Y-W);
  }
double CashCarp(void dYdX(double,double*,
    double*),double*X,double dX,double*Y)
  {
  double k1,k2,k3,k4,k5,k6,W;
  dYdX(*X,Y,&k1);
  W=*Y+k1*dX/5.;
  dYdX(*X+dX/5.,&W,&k2);
  W=*Y+(3.*k1+9.*k2)*dX/40.;
  dYdX(*X+12.*dX/40.,&W,&k3);
  W=*Y+(3.*k1-9.*k2+12.*k3)*dX/10.;
  dYdX(*X+6.*dX/10.,&W,&k4);
  W=*Y+(-11.*k1+135.*k2-140.*k3+70.*k4)*dX/54.;
  dYdX(*X+dX,&W,&k5);
  W=*Y+(3262.*k1+37800.*k2+4600.*k3
    +44275*k4+6831.*k5)*dX/110592.;
```

```
dYdX(*X+7.*dX/8.,&W,&k6);
W=*Y+(9361.*k1+38500.*k3+20125.*k4
  +27648.*k6)*dX/95634.;
*Y=*Y+(2808050.*k1+10550600.*k3+6721925.*k4
  +531009.*k5+6870528.*k6)*dX/27482112.;
*X=*X+dX;
return(*Y-W);
}
```

We will use the first example problem in Chapter 22 for a test case (Equation 22.1). The code can be found in the on-line archive in file embedded.c. The results can be found in embedded.xls. Since we know the exact solution (Equation 22.2), we can calculate the exact error and will just consider the absolute values. We solve the same problem for 4 different step sizes (h=0.05, 0.10, 0.15, and 0.20). The exact error (E2-E1 in the following code) is equal to the estimated change in the dependent variable (i.e., the result of applying the Runge-Kutta method for a single step, E2=Y2-Y1) minus the actual change in the dependent variable (i.e., the difference in the analytical solution for the same step in the independent variable, E1=f2-f1). We can write a single function to handle all four methods:

```
void testMethod(double method(void dYdX(double,double*,
  double*),double*,double,double*),double dX)
{
double dY1,dY2,E1,E2,f1,f2,X,Y1,Y2;
X=Y2=f2=0.;
do{
  f1=f2;
  Y1=Y2;
  E2=method(dYdX,&X,dX,&Y2);
  f2=f(X);
  dY1=f2-f1;
  dY2=Y2-Y1;
  E1=dY2-dY1;
  }while(X<3.);
}
```

We can use the preceding function to test all four methods at four different step sizes:

```
for(dX=0.05;dX<0.21;dX+=0.05)
  {
  testMethod(HuenEuler,dX);
  testMethod(Fehlberg,dX);
  testMethod(BogackiShampine,dX);
  testMethod(CashCarp,dX);
  }
```

Results for the Huen-Euler method are shown in the following figure:

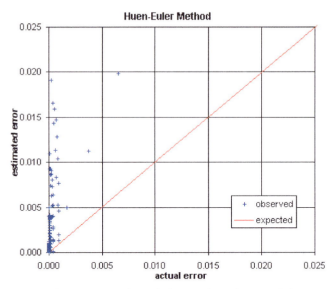

The estimated error is consistently larger than the actual (i.e., most of the blue +s are above the red line). There is also no pattern to the estimated error with respect to the actual error. Therefore, this method is useless—at least in this particular case. This next figure shows the same thing only for the Bogacki-Shampine method:

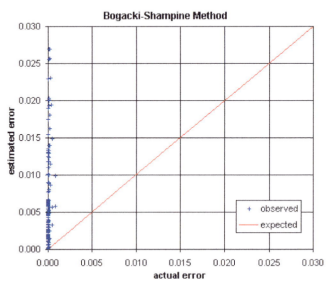

The estimated error is consistently larger than the actual (i.e., most of the blue +s are above the red line). Again, there is no pattern. This method is also useless—at least in this particular case. This next figure shows the Fehlberg:

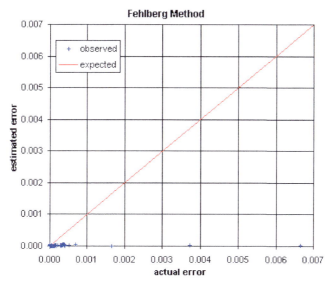

This time the estimated error is consistently smaller than the actual (i.e., most of the blue +s are below the red line). Still, there is also no pattern. This method is also useless—in this particular case. Next the Cash-Carp method:

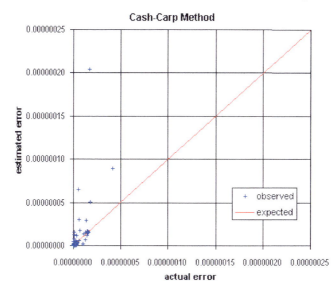

Finally, we start to see a pattern! An exploded view of the lower left corner of this figure shows the pattern in more detail:

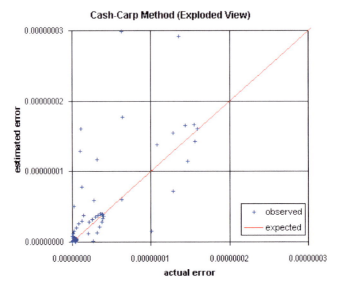

We could probably "fix" the first three methods, but what would be the point? This entire process has theoretical value, but is of little practical use. It might have been important before the invention of computers capable of performing billions of instructions per second, but that was so many nanoseconds ago that I've lost count.

Instead of applying a double method that yields an estimate of the error in twice as many microseconds, why not just cut the step size in half? If you don't get the same answer, cut it in half again. We've already presented code that calls a function, that calls another function, that implements whatever method from a table. Just put the whole thing inside a function and try smaller steps until it converges to your satisfaction.

Chapter 26. Finite Difference Method

Marching methods such as Runge-Kutta work well enough for initial value problems and some very simple boundary value problems, but they don't work for most boundary value problems, especially when the domain is complex. Finite difference methods work well for simple geometries. For complex geometries, finite element methods are preferable. We will cover those in Chapter 27.

Steady-State Heat Conduction

You rarely see the transient heat conduction equation with variable properties written out in its full form. In Cartesian coordinates (x,y,z) this is:

$$\frac{\partial(\rho C T)}{\partial t} = \frac{\partial}{\partial x}\left(k\frac{\partial T}{\partial x}\right) + \frac{\partial}{\partial y}\left(k\frac{\partial T}{\partial y}\right) + \frac{\partial}{\partial z}\left(k\frac{\partial T}{\partial z}\right) \quad (26.1)$$

Most often, the density (ρ), specific heat (C), and thermal conductivity (k) are presumed constant and taken outside the differential. In this first finite difference problem, we will consider steady-state (i.e., time-independent) conduction through a bar having variable thermal conductivity. In this case, Equation 25.1 reduces to:

$$0 = \frac{\partial}{\partial x}\left(k\frac{\partial T}{\partial x}\right) \quad (26.2)$$

As this is a second-order differential equation, we must consider at least three points, representing $x-\Delta x$, x, and $x+\Delta x$. We also need to know the thermal conductivity at $x-\Delta x/2$ and at $x+\Delta x/2$ in order to complete the expression:

$$0 = \frac{\left(k_{x+\frac{\Delta x}{2}}\frac{T_{x+\Delta x}-T_x}{\Delta x}\right) - \left(k_{x-\frac{\Delta x}{2}}\frac{T_x-T_{x-\Delta x}}{\Delta x}\right)}{\Delta x} \quad (26.3)$$

If Δx is constant, this reduces to:

$$k_{x+\frac{\Delta x}{2}}(T_{x+\Delta x}-T_x) = k_{x-\frac{\Delta x}{2}}(T_x-T_{x-\Delta x}) \quad (26.4)$$

Equation 26.4 can be solved for T_x:

$$T_x = \frac{k_{x+\frac{\Delta x}{2}}T_{x+\Delta x} + k_{x-\frac{\Delta x}{2}}T_{x-\Delta x}}{k_{x+\frac{\Delta x}{2}} + k_{x-\frac{\Delta x}{2}}} \quad (26.5)$$

We can easily construct an Excel® spreadsheet (finite_difference1.xls) to solve this problem. Because the temperatures depend on the thermal conductivities and the thermal conductivity depends on the temperature, this is an implicit (and nonlinear) calculation. Excel® can be programmed to handle this automatically (set tools options calculations iteration ON). For this example, we can define the thermal conductivity in a table (cells L3:M13) and use an interpolation macro (included) to find the value. We define the temperature at the end points (cells B2 and B12) and Excel® does the rest. We must help the iterations get started, so we use the IFERROR() function to assume an initial value. The result is:

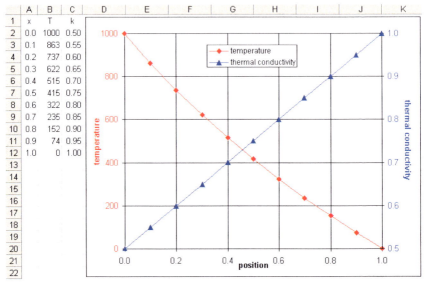

If the thermal conductivity were a constant, the blue curve would be flat and the red curve would be a straight line between the two end points. The formula in cell B3 is =IFERROR((B2*(C3+C4)+B4*(C2+C3))/(C2+2*C3+C4),500), which is Equation 26.5. We can extend this same finite difference to a second dimension, y, and solve for the corresponding result:

$$T_{xy} = \frac{\Delta y \left(k_{x+\frac{\Delta x}{2}} T_{x+\Delta x} + k_{x-\frac{\Delta x}{2}} T_{x-\Delta x} \right) + \Delta x \left(k_{y+\frac{\Delta y}{2}} T_{y+\Delta y} + k_{y-\frac{\Delta y}{2}} T_{y-\Delta y} \right)}{+ \Delta y \left(k_{x+\frac{\Delta x}{2}} + k_{x-\frac{\Delta x}{2}} \right) + \Delta x \left(k_{y+\frac{\Delta y}{2}} + k_{y-\frac{\Delta y}{2}} \right)}$$

(26.6)

On the 2D tab of this same spreadsheet there is a two-dimensional problem with a rectangular domain having $\Delta x = \Delta y$ and prescribed temperatures along each of the 4 boundaries. It's also iterative and uses the same lookup table for thermal conductivity. You can change the temperatures along any of the boundaries or the thermal conductivities in the table and the solution will automatically adjust. Note that the corner cells in the temperature and thermal conductivity array are all empty, as these aren't used for anything and don't enter into the calculations. The result is:

	A	B	C	D	E	F	G	H	I	J	K	L	M
1						Temperatures							
2		1000	900	800	700	600	500	400	300	200	100	0	
3	1000	911	816	722	629	537	446	357	270	185	105	37	0
4	950	847	751	661	573	489	406	325	248	174	106	47	0
5	900	794	699	612	529	450	374	301	231	165	103	49	0
6	850	746	654	571	493	419	348	281	217	156	99	48	0
7	800	701	614	536	462	393	327	264	204	148	95	46	0
8	750	659	578	505	436	371	309	250	193	140	90	44	0
9	700	618	545	477	413	352	293	237	184	133	86	42	0
10	650	579	514	453	393	335	280	226	175	127	82	39	0
11	600	543	487	432	376	322	269	217	167	120	77	37	0
12	550	511	466	415	363	311	260	209	160	114	71	32	0
13	500	490	452	405	355	304	253	203	154	107	62	23	0
14		500	450	400	350	300	250	200	150	100	50	0	
15						Thermal Conductivities							
16		0.50	0.54	0.57	0.62	0.66	0.71	0.76	0.81	0.87	0.93	1.00	
17	0.50	0.53	0.57	0.61	0.65	0.69	0.73	0.78	0.83	0.88	0.93	0.98	1.00
18	0.52	0.56	0.59	0.63	0.67	0.71	0.75	0.80	0.84	0.89	0.93	0.97	1.00
19	0.54	0.58	0.62	0.65	0.69	0.73	0.77	0.81	0.85	0.89	0.93	0.97	1.00
20	0.56	0.60	0.64	0.67	0.71	0.75	0.79	0.82	0.86	0.90	0.93	0.97	1.00
21	0.57	0.62	0.65	0.69	0.73	0.76	0.80	0.83	0.87	0.90	0.94	0.97	1.00
22	0.59	0.63	0.67	0.70	0.74	0.77	0.81	0.84	0.87	0.91	0.94	0.97	1.00
23	0.62	0.65	0.69	0.72	0.75	0.78	0.82	0.85	0.88	0.91	0.94	0.97	1.00
24	0.64	0.67	0.70	0.73	0.76	0.79	0.82	0.86	0.89	0.92	0.95	0.97	1.00
25	0.66	0.69	0.71	0.74	0.77	0.80	0.83	0.86	0.89	0.92	0.95	0.98	1.00
26	0.68	0.70	0.72	0.75	0.78	0.81	0.84	0.87	0.90	0.92	0.95	0.98	1.00
27	0.71	0.71	0.73	0.76	0.78	0.81	0.84	0.87	0.90	0.93	0.96	0.98	1.00
28		0.71	0.73	0.76	0.79	0.81	0.84	0.87	0.90	0.93	0.97	1.00	

The temperatures in the rose cells are calculated using Equation 26.6 and the thermal conductivities in the sky blue cells are calculated from these using the interpolation macro. Once again, we must use the IFERROR() function to help Excel® get things started. If we were writing a code to do this, we would simply initialize all the variables to some reasonable value before starting to iterate for a solution.

Notice that cells A3 and B2 are equal, as are L2 and M3, A13 and B14, L14 and M13. These don't have to be equal in this spreadsheet, but in our finite difference scheme, these corner points represent the same location in space. In the real world, you don't have temperature discontinuities. To be realistic, these should be equal.

The above temperature map is illustrated below:

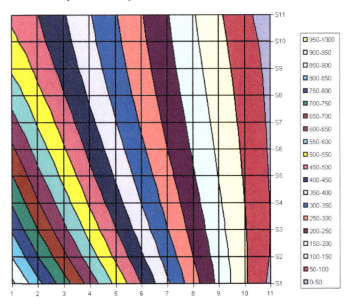

and here:

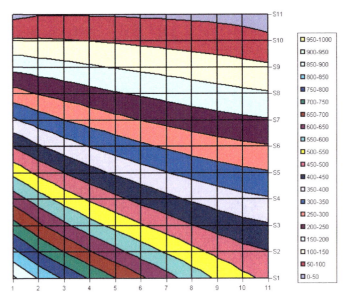

These are both sort of upside down and backwards because that's the way Excel® does it. The 2D graphs provided by Excel® leave much to be desired.

To produce a quality 2D graph, you'll need a tool like Tecplot® or TP2. The latter is available free at the following link:

http://www.dudleybenton.altervista.org/software/index.html

TP2 produces the following proper rendering of this temperature field:

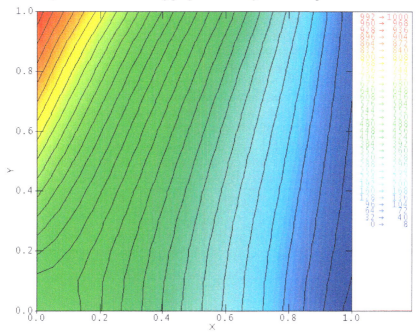

More details on finite difference operators may be found in Appendix H.

Transient Heat Conduction

No discussion of finite difference methods would be complete without mention of the Crank-Nicholson technique. For one-dimensional transient conduction with approximately constant properties Equation 26.1 simplifies to:

$$\rho C \frac{\partial T}{\partial t} = k \frac{\partial^2 T}{\partial x^2} \qquad (26.7)$$

We might express this in terms of finite differences:

$$\rho C \left(\frac{T_{t+\Delta t,x} - T_{t,x}}{\Delta t} \right) = k \left(\frac{T_{t,x+\Delta x} - 2T_{t,x} + T_{t,x-\Delta x}}{\Delta x^2} \right) \qquad (26.8)$$

Equation 26.8 is a central difference in space and a fully explicit (forward Euler) difference in time. It would be more accurate and also more stable to evaluate the right-hand side of Equation 26.8 at time t and also at $t+\Delta t$, using the average to compute the temporal derivative. This would be an implicit difference in time, often called the *midpoint* method. This requires knowledge of the temperatures before and after the time step in order to calculate each step through time.

If we had planned to solve a matrix for the temperatures at the next step, this merely adds a few more terms, which can be rearrange and folded into the calculations we were already going to perform. This is the Crank-Nicholson method and is quite stable, even for large time steps. While this idea is very clever and quite effective, it has little practical use for several reasons. First, you rarely want to use the kind of algorithms required to solve this type of matrix. Second, Crank-Nicholson is fairly easy to encode for a 1D problem, but it's a real hassle for 2D and beyond. Who needs to solve 1D transient heat conduction problems? Third, you can just march through time using a Runge-Kutta method and forget about solving a matrix.

Because the Crank-Nicholson method is interesting theoretically and historically significant, we will set up and solve a simple problem in one dimension with 4 nodes. The 4 nodal point equations become:

$1+\beta$	$-\beta/2$	0	0		$T_{1,t+\Delta t}$		$1-\beta$	$\beta/2$	0	0		$T_{1,t}$
$-\beta/2$	$1+\beta$	$-\beta/2$	0	×	$T_{2,t+\Delta t}$	=	$\beta/2$	$1-\beta$	$\beta/2$	0	×	$T_{2,t}$
0	$-\beta/2$	$1+\beta$	$-\beta/2$		$T_{3,t+\Delta t}$		0	$\beta/2$	$1-\beta$	$\beta/2$		$T_{3,t}$
0	0	$-\beta$	$1+\beta$		$T_{4,t+\Delta t}$		0	0	β	$1-\beta$		$T_{4,t}$

The parameter β is equal to:

$$\beta = \frac{k\Delta t}{\rho C \Delta x^2} \qquad (26.10)$$

The entire process can be implemented in an Excel® spreadsheet (Crank-Nicholson.xls in the on-line archive) using the matrix multiply MMULT() and matrix inversion MINVERSE() functions. The matrices and properties are:

Crank-Nicholson Matrices								properties	
matrix A				matrix B				Δt	1.00
2.0	-0.5	0.0	0.0	0.0	0.5	0.0	0.0	Δx	1.00
-0.5	2.0	-0.5	0.0	0.5	0.0	0.5	0.0	ρ	1.00
0.0	-0.5	2.0	-0.5	0.0	0.5	0.0	0.5	C	1.00
0.0	0.0	-1.0	2.0	0.0	0.0	1.0	0.0	k	1.00
calculated parameter, $\beta=k\Delta t/(\rho C\Delta x^2)$									1.00

Change the numbers in bold blue and the results will automatically update.

The temperatures are shown in this next figure:

Crank-Nicholson time steps
calculated temperatures at each time step

t=0	t=Δt	t=2Δt	t=3Δt	t=4Δt	t=5Δt	t=6Δt	t=7Δt	t=8Δt
100	46.4	38.0	30.8	26.3	22.5	19.3	16.5	14.2
100	85.6	66.3	56.7	48.3	41.5	35.6	30.6	26.2
100	95.9	85.1	73.2	63.1	54.1	46.5	39.9	34.3
100	97.9	90.5	79.1	68.1	58.6	50.3	43.2	37.1

The initial temperatures are in the first column (all set to 100). Each time step to the right is an application of the matrix operations in Equation 26.9. The same thing could be implemented with a marching technique, the simplest being:

explicit marching scheme

0	0	0	0	0	0	0	0	0
100	75.0	56.3	42.2	31.6	23.7	17.8	13.3	10.0
100	100.0	87.5	78.1	66.9	57.1	47.9	40.0	33.0
100	100.0	100.0	91.7	82.6	72.2	62.1	52.7	44.2
100	100.0	95.8	87.2	77.4	67.2	57.4	48.4	40.5

d^2/dx^2

-50	-37.5	-28.1	-21.1	-15.8	-11.9	-8.9	-6.7	-5.0
0	-25.0	-18.8	-22.4	-19.6	-18.4	-15.9	-13.9	-11.8
0	0.0	-16.7	-18.1	-20.9	-20.0	-18.9	-16.9	-14.9
0	-8.33	-17.4	-19.5	-20.5	-19.5	-17.9	-15.9	-13.9

Open Channel Flow

The conservation of mass and momentum for unsteady flow in an open channel is described by the following two partial differential equations, respectively:

$$\frac{\partial A}{\partial t} + \frac{\partial Q}{\partial x} + q = 0 \qquad (26.11)$$

$$\frac{\partial Q}{\partial t} + gA\frac{\partial h}{\partial x} + \frac{1}{A}\frac{\partial Q^2}{\partial x} = gA(S_0 - S_f) \qquad (26.12)$$

Here, Q is the flow (or discharge, as it is often called in open channel flow), A is the cross sectional area, q is the lateral inflow (not contributing momentum along the channel), g is the acceleration of gravity, h is the depth (or water surface elevation), x is the distance along the channel, t is time, S_f is the friction slope, and S_0 is the bottom slope (positive with decline in the downstream direction). Equation 26.12 is called the De-Saint Venant Equation.

The slope (geometric and effective arising from friction along the channel bottom) for open channel flow is empirically described by Manning's Equation:

$$Q = VA = \left(\frac{1.49}{n}\right) AR^{\frac{2}{3}} \sqrt{S} \qquad (26.13)$$

In the preceding equation V is the bulk velocity, A is the cross-sectional area, R is the hydraulic radius, S is the geometric slope, and n is Manning's factor. The factor 1.49 arises from the unit conversion for meters to feet $3.2808^{\frac{1}{3}}$. This is unity when using SI units.

We will solve the conservation of mass and momentum equations simultaneously using a finite difference approach. The more interesting aspect of this example is that of sloshing, which is something water often does, especially when you open one dam and shut another. Sloshing makes this a *stiff* system of equations, meaning the derivatives with respect to time can easily run away with overshoot.

We will use two techniques to handle the problem of sloshing: 1) predictor/corrector and 2) MacCormack spatial derivatives. The former is like iteratively approximating the midpoint method discussed in the previous example. The latter has an effect of reducing sloshing like *numerical viscosity*.

The simple predictor corrector method is easily implemented by initializing the new values of flow and elevation equal to the old and then successively substituting revised values several times before advancing time to the next step. In this case experience has shown that 4 iterations is adequate. Partial differentials for the finite difference equations are calculated by the following function:

```
void Differentials(double*Q,double*H,double*dQdt,
   double*dHdt)
{
int i;
double Bavg,dHdx,dQdx,dQQAdx;
for(i=0;i<sections;i++)
   QQA[i]=Q[i]*Q[i]/A[i];
for(i=0;i<sections-1;i++)
   {
   dQdx=(Q[i+1]-Q[i])/dX;
   Bavg=(B[i]+B[i+1])/2.;
   dHdt[i]=(Qinf[i]-dQdx)/Bavg;
   }
for(i=1;i<sections-1;i++)
   {
   dHdx=(H[i]-H[i-1])/dX;
   dQQAdx=(QQA[i+1]-QQA[i-1])/2./dX;
   dQdt[i]=-gravity*A[i]*(dHdx+Sf[i])-dQQAdx;
   }
```

The step forward in time is accomplished by the following code with *ncorr* corrector iterations:

```
/* corrector: iterate ncor times */
        for(iter=0;iter<ncorr;iter++)
            {
/* update slopes */
        for(i=1;i<sections-1;i++)
            Sf[i]=So[i]*Q[new][i]*fabs(Q[new][i]);
/* calculate differentials (pass new arrays) */
        Differentials(&Q[new][0],&H[new][0],
    &dQdt[new][0],&dHdt[new][0]);
/* step forward in time */
        for(i=0;i<sections-1;i++)
            {
            H[new][i]=H[old][i]+dt*
    (dHdt[old][i]+dHdt[new][i])/2.;
            H[new][i]=max(Hmin,min(Hmax,H[new][i]));
            }
        for(i=1;i<sections-1;i++)
            Q[new][i]=Q[old][i]+dt*
    (dQdt[old][i]+dQdt[new][i])/2.;
            }
        hour+=1./nstep;
```

The predictor/corrector step is initialized by the following code:

```
/* predictor: set Hnew and Qnew to Hold and Qold */
        for(i=0;i<sections-1;i++)
            {
            H[new][i]=H[old][i]+dt*dHdt[old][i]/2.;
            H[new][i]=max(Hmin,min(Hmax,H[new][i]));
            }
        for(i=1;i<sections-1;i++)
            Q[new][i]=Q[old][i]+dt*dQdt[old][i]/2.;
```

The MacCormack upwind/downwind differencing scheme is accomplished by storing the elevation, *H*, and flow, *Q*, in a two-dimensional arrays:

```
    double H[2][sections-1];/* water surface elevation */
    double dHdt[2][sections-1];
    double dQdt[2][sections];
    double Q[2][sections]; /* flow */
```

and the rest of the variables in one-dimensional arrays:

```
    double A[sections]; /* area */
    double B[sections]; /* width */
    double Havg[sections]; /* average elevation */
    double Qinf[sections-1]; /* local inflow */
    double QQA[sections]; /* Q²/A */
    double Sf[sections-1]; /* friction slope */
    double So[sections]; /* total slope */
```

The spatial finite differences are flip-flopped every time step by swapping the integer index *new* and *old*.

```
/* swap new/old index for MacCormack step */
new=1-new;
old=1-new;
```

The differencing function is called with the index of the arrays so that this changes with each step, but doesn't require any additional code to implement beyond this:

```
/* calculate differentials (pass new arrays) */
Differentials(&Q[new][0],&H[new][0],&dQdt[new][0],
    &dHdt[new][0]);
```

The function receives a pointer to the first element in either the new or old section of the array, respectively:

```
void Differentials(double*Q,double*H,double*dQdt,
    double*dHdt)
```

In order to get the calculations started, two initialization days are run with the same inputs before the actual calculations:

```
    if(initialize)
      {
/* first time through take two days to initialize
    everything */
     ndays=3;
     hour=-24.*(ndays-1);
     for(i=0;i<sections-1;i++)
        H[new][i]=hdn+(hup-hdn)*pow((sections-2.-
    i)/(sections-2.),4);
     for(i=0;i<sections;i++)
        Q[new][i]=Qups[0]+(Qdns[0]-Qups[0])*i/(sections-
    1.);
      }
else
   initialize=FALSE;
```

The runoff (drainage into the reservoir from rain) and local inflow from a side stream are added before the iteration begins:

```
/* add runoff to each cell */
    for(i=0;i<sections-1;i++)
       Qinf[i]=Runoff;
/* add inflow from stream */
    Qinf[infsection]+=Qstream/dX;
```

The geometry (area, width, and slope) are initialized using the predictor values:

```
        Havg[0]=H[new][0];
        for(i=1;i<sections;i++)
           Havg[i]=(H[new][i-1]+H[new][i])/2.;
        Geometry(Havg,A,B,So);
```

The input is supplied in 24-hour blocks, read from a file. Typical input is:

```
680 680 0 0 <-- WATTS BAR TAILWATER ELEV./CHICKAMAUGA
   HEADWATER ELEV./
 1     0      0              TOTAL LOCAL INFLOW/HIWASSEE
   RIVER INFLOW
 2     0      0
 3     0      0 <-- HOUR/WATTS BAR, CHICKAMAUGA RELEASE
   CFS
 4 10000  10000
 5 10000  10000
 6 15000  10000
 7  5000  20000
 8  5000  20000
 9  5000  30000
10 15000  30000
11 20000  30000
12 30000  30000
13 30000  30000
14 40000  30000
15 40000  30000
16 40000  30000
17 40000  30000
18 40000  30000
19 40000  30000
20 20000      0
21 10000      0
22     0      0
23     0      0
24     0      0
```

This particular geometry is for the Chickamauga Reservoir on the Tennessee River between Watts Bar Dam and Chickamauga Dam. The side stream entering the channel is the Hiawassee River. The midsection where flow and elevation are of special interest is at Soddy Daisy, the site of Sequoyah Nuclear Plant. All of the associated files can be found in the on-line archive named SQFLOW.*. The following is typical output of the model:

```
Chickamauga One-Dimensional Reservoir Routing Model
using multi-step predictor-corrector method
initializing day 1
initializing day 2
routing day 1
hour  Qups    Qmid  Qdns   Hups   Hmid   Hdns
 100     0  -14567     0 680.33 680.12 680.18
 200     0  -15484     0 680.16 680.01 680.00
 300     0    3186     0 680.23 679.98 679.87
 400 10000   18744 10000 680.73 679.99 679.97
 500 10000   17710 10000 680.80 680.02 680.04
 600 15000    8011 10000 680.92 680.05 680.05
 700  5000   16007 20000 680.14 680.06 679.98
```

```
 800   5000  22859  20000  679.81  679.99  679.99
 900   5000  21634  30000  679.96  679.97  679.89
1000  15000  22959  30000  680.83  679.85  679.83
1100  20000  17727  30000  681.28  679.77  679.73
1200  30000  23909  30000  682.12  679.70  679.62
1300  30000  32524  30000  682.49  679.69  679.60
1400  40000  32546  30000  683.54  679.69  679.66
1500  40000  29620  30000  684.06  679.71  679.65
1600  40000  30575  30000  684.31  679.71  679.66
1700  40000  31543  30000  684.45  679.73  679.66
1800  40000  33243  30000  684.54  679.75  679.69
1900  40000  32784  30000  684.61  679.77  679.72
2000  20000  22324      0  682.64  679.91  679.91
2100  10000   4075      0  680.90  680.12  680.18
2200      0   5655      0  679.61  680.18  680.23
2300      0   7815      0  679.39  680.22  680.21
2400      0  -1130      0  679.94  680.23  680.26
```

A graph of the results is shown in this next figure:

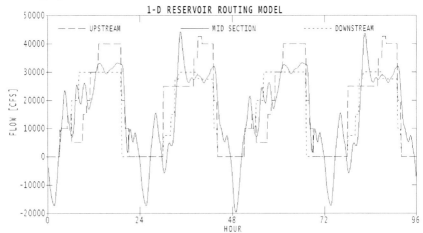

The upstream and downstream flows have a step shape, because this is the way the dams are operated: gates are either open or closed and there are a fixed number of open settings for each gate. Flow at the midpoint exhibits significant sloshing. In fact, the sloshing is so pronounced that 49 hours into this cycle, the flow is back upstream at a rate of 20,000 cubic feet per second (566m^3/s).

When you close the downstream dam, it sends a wave back up stream. The midsection Sequoyah shown by the solid line is 12.5 miles (20.1 km) upstream of Watts Bar dam. These *waves* travel all the back to the upstream dam 58.6 miles (94.3 km) and are reflected back, resulting in smaller highs and lows in the flow at subsequent times.

Open Channel Flow with Heat Transfer and Dry Bed

The preceding model will only work with a fully-wetted, mostly full channel. It also doesn't consider thermal energy or temperature. We will now modify the equations to handle the case when all the water runs out, leaving the bed dry. We will also handle heat transfer so that the temperature may be calculated. In the preceding model, both the upstream and downstream were regulated by dams. In this next case, the upstream will be controlled by a dam, but the downstream is free to discharge into a larger body of water, so that it has no rigid boundary condition.

The seminal reference on kinematic waves is the USGS monograph by Miller.[40] This work (which is readily available on-line) explains the development of the governing equations as well as various solutions and has been the principal guidance for surface water models ever since. The kinematic wave equation is a special case of Equation 26.11.

$$\frac{1}{c}\frac{\partial Q}{\partial t} + \frac{\partial Q}{\partial x} = 0 \qquad (26.14)$$

The friction term (i.e., right-hand side of Equation 26.11) has disappeared, as momentum is presumed dominant. The new term, c, is the celerity (i.e., wave propagation speed). As it turns out, we will need to add the friction term back in order to better fit observed data. At this point of Miller's development, the celerity is rather vague, perhaps equal to dQ/dA, but more likely $c=\sqrt{gh}$.

After a discussion of the method of characteristics, wave propagation and reflection, Miller presents the following finite difference equation:

$$Q_{x+\Delta x, t+\Delta t} = C_1 Q_{x,t} + C_2 Q_{x,t} + \Delta t + C_3 Q_{x+\Delta x, t} \qquad (26.15)$$

Equation 26.15 derives from the Muskingum-Cung method using variable parameters C_1, C_2, and C_3 as described by Ponce and Yevjevich.[41] The parameters are:

$$C_1 = \frac{1+C-D}{1+C+D}$$
$$C_2 = \frac{-1+C+D}{1+C+D} \qquad (26.16)$$
$$C_3 = \frac{1-C+D}{1+C+D}$$

---

[40] Miller, J. E., "Basic Concepts of Kinematic-Wave Models," U. S. Geological Survey Professional Paper No. 1302, 1984.

[41] Ponce, V. M. and Yevjevch, V., "Muskingum-Cunge Method with Variable Parameters," *Journal of the Hydraulics Division* of the American Society of Civil Engineers, Vol. 104, HY 3, pp. 353-360, 1978.

In Equation 26.16 the parameter C is the Courant number and D is a dimensionless term that brings the friction factor back into play.

$$C = \frac{c\Delta t}{\Delta x}$$
$$D = \frac{q}{S_0 c \Delta x} \qquad (26.17)$$

As with the previous routing model, we must initialize the calculations in order to get the process started:

```
/* initialize flows and celerrities */
  if(initialize)
    {
    initialize=FALSE;
    qinit1=400.;
    dx=12.5*5280./(sections-1);

    v=qinit1/B[0]/pow(qinit1*rn[0]/(B[0]*pow(So[0],0.5)*
    1.486),0.6);
    hnew=qinit1/v/B[0];
    for(i=0;i<sections;i++)
       {
       Q[new][i]=qinit1;
       H[new][i]=hnew;
       C[new][i]=1.5*v;
       Triv[new][i]=Tdam;
       Tbed[new][i]=Tdam;
       rvbed[i]=dbed*rhobed*B[i]*dx;
       }
    }
```

Then we swap the old and new index and iterate (predictor/corrector) to calculate the celerity because it's implicit and not simply equal to $\sqrt{gh}$.

```
/* switch old/new index */
  while(TRUE)
    {
    new=1-new;
    old=1-new;
/* approximate new values by old values */
    for(i=0;i<sections;i++)
       {
       Q[new][i]=Q[old][i];
       H[new][i]=H[old][i];
       C[new][i]=C[old][i];
       Triv[new][i]=Triv[old][i];
       }
/* determine maximum celerity */
    iter1=0;
    cmx=0.;
```

```
      while(TRUE)
        {
        iter1+=1;
        if(iter1>niter1)
          goto l_100;
        cmax=C[new][0];
        for(i=0;i<sections;i++)
            cmax=max(cmax,C[new][i]);
        Q[new][0]=Qdam;
        Triv[new][0]=Tdam;
/* if initial flow is not zero,check it with next
   boundary condition. whichever is greater must be
   used to calculate to retain stability */
        if(Q[new][0]>0.)
            C[new][0]=1.5*Q[new][0]/(B[0]*
   (Q[new][0]*rn[0]/(B[0]*pow(So[0],0.5)*1.486))*0.6);
        cmax=max(cmax,C[new][0]);
```

We must limit the step size in order to not exceed the Courant condition (i.e., disturbances can't propagate faster than $\Delta x/\Delta t$). We iterate for flow after the time step (predictor/corrector method again):

```
/* calculate step size and put an upper limit on dt */
        dt=min(dt0,dx/cmax);
/* start of loop that calculates the q for each section
   of the river */
        for(i=1;i<sections;i++)
          {
          iter2=0;
          while(TRUE)
            {
            iter2+=1;
/* check for non-convergence in flow */
            if(iter2>niter2)
              goto l_130;
            qbar=(Q[old][i-1]+Q[old][i])/2.;
            cbar=(C[new][i-1]+C[new][i]+C[old][i-
   1]+C[old][i])*0.25;
            qb=(Q[new][i-1]+Q[new][i])/2.;
            qbxt=(qbar+qb)/2.;
            theta=max(0.,min(0.9999,(1.-
   qbxt/(B[i]*So[i]*cbar*dx))/2.));
            c11=dx/(cbar*dt);
            c3=exp(-cbar*dt/(dx*(1.-theta)));
            c1=1.-c11*(1.-c3);
            c2=c11*(1.-c3)-c3;
            qnew=Q[new][i];
            Q[new][i]=c1*Q[new][i-1]+c2*Q[old][i-
   1]+c3*Q[old][i];
            if(Q[new][i-
   1]>Q[new][i]&&Q[old][i]>Q[new][i])
```

```
        Q[new][i]=Q[old][i];
     change=fabs(qnew-Q[new][i]);
     if(iter2>6)
        Q[new][i]=(Q[new][i]+qnew)/2.;
     if(iter2>1&&change<50.)
        break;
     }
  }
```

Then update the geometry, slopes, and depths, recalculate the friction factor, and correct the celerity—all of this inside an iteration loop.

```
/* calculate new slopes and depths */
     for(i=0;i<sections;i++)
        {
        v=0.1;
        if(Q[new][i]>0.)
           {
/* s is the friction slope term. put a lower limit on
   it so that the stage calculation will not have
   division by zero or sqrt of a negative number */
           s=max(0.000125,(Q[new][i]-
   Q[old][i])/(dt*B[i]*sq(C[new][i])+So[i]));
/* this is the stage rating curve */
           H[new][i]=pow(Q[new][i]*rn[i]
     /(1.486*B[i]*sqrt(s)),0.6);
           }
/* calculate velocity,celerity,flow */
           v=Q[new][i]/(B[i]*H[new][i]);
           C[new][i]=1.5*v;
           }
        change=fabs(cmax-cmx);
        if(iter1>1&&change<0.2)
           break;
        cmx=cmax;
        }
```

We calculate the temperature by the conservation of energy. The rate of increase in energy of the water in a cell is equal to that carried in by the water entering the cell minus that carried out by the water exiting the cell plus the heat transfer from the atmosphere to the water in the cell plus the heat transfer from the streambed into the water in the cell. The energy of any element of water is equal to the volume times the density times the heat capacity times the temperature ($E=V\rho CT$).

The rate of heat exchange with the air or streambed is equal to the area times the heat transfer coefficient times the temperature difference (heat transfer=$Ah\Delta T$). The symbols can become a bit confused here, as Q is often used to denote flow and heat transfer and h is often used to denote height and heat transfer coefficient. The calculation of temperature begins thus:

```
/* determine water temperature by conservation of
   energy (first law of thermodynamics) */
    for(i=1;i<sections;i++)
       {
       rvold=rho*H[old][i]*B[i]*dx;
       rvnew=rho*(Q[old][i-1]+Q[new][i-1]-Q[old][i]-
  Q[new][i])*dt/2.+rvold;
/* determine if bed is dry */
       drybed=FALSE;
       if(rvnew<1000.)
          drybed=TRUE;
/* determine energy in a section */
       ein=rho*dt*(Q[old][i-1]*Triv[old][i-1]+Q[new][i-
  1]*Triv[new][i-1])/2.;

       eout=rho*dt*(Q[old][i]*Triv[old][i]+Q[new][i]*Triv[n
       ew][i])/2.;
       eold=Triv[old][i]*rvold;
/* set bounds on temperature to assure agreement with
   the second law of thermodynamics */
       tmin=min(Triv[old][i-1],Triv[old][i]);
       tmin=min(tmin,Triv[new][i-1]);
       tmin=min(tmin,T0);
       tmin=min(tmin,Tbed[old][i]);
       tmax=max(Triv[old][i-1],Triv[old][i]);
       tmax=max(tmax,Triv[new][i-1]);
       tmax=max(tmax,T0);
       tmax=max(tmax,Tbed[old][i]);
/* iterate to determine new temperature */
       iter3=0;
       while(TRUE)
          {
          iter3+=1;
/* check for non-convergence */
          if(iter3>niter3)
             goto l_120;
```

There is no advection, only convection, when the streambed is dry, so we must handle both cases:

```
/* skip river temperature calculations if bed is dry */
       change=0.;
       if(!drybed)
          {
/* calculate net thermal advection */
          thad=edifft*(B[i-1]*H[old][i-1]
     *(Triv[old][i-1]-Triv[old][i])+B[i]*H[old][i]
     *(Triv[old][min(sections,i+1)-1]-Triv[old][i])
     +B[i-1]*H[new][i-1]*(Triv[new][i-1]-Triv[new][i])
     +B[i]*H[new][i]*(Triv[old][min(sections,i+1)-1]
     -Triv[new][i]))/2./dx;
```

```
    /* calculate surface heat transfer */
            surfht=dx*B[i]*htcsur*(2.*T0-Triv[new][i]
    -Triv[old][i])/2./3600.;
            }
    /* estimate the heat transfer coefficient between the
       river and the bed in a section and then calculate
       heat transfer to the river bed */
         if(drybed)
             {
             htcbed=htcsur;
             rbedht=dx*B[i]*htcbed
    *(Tbed[new][i]+Tbed[old][i]-2.*T0)/2./3600.;
             }
         else
             {
             htcbed=2./(H[new][i]/kwater+(dbed/kbed));
             rbedht=dx*B[i]*htcbed
    *(Tbed[new][i]+Tbed[old][i]-Triv[new][i]
    -Triv[old][i])/2./3600.;
             }
    /* calculate new energy in a section of the river */
            enew=eold+ein-eout+(surfht+thad+rbedht)*dt;
            if(!drybed)
               tnew=enew/rvnew;
            tnew=max(tmin,min(tmax,tnew));
            change=fabs(Triv[new][i]-tnew);
            Triv[new][i]=(Triv[new][i]+tnew)/2.;
    /* calculate new river bed temperature in a section */
            tnew=Tbed[old][i]-rbedht*dt/rvbed[i]/cbed;
            tnew=max(tmin,min(tmax,tnew));
            change=max(change,fabs(Tbed[new][i]-tnew));
            Tbed[new][i]=(Tbed[new][i]+tnew)/2.;
            if(drybed)
               Triv[new][i]=Tbed[new][i];
    /* check for convergence in temperature */
            if(iter3>1&&change<1.)
               break;
            }
        }
```

We then update time and check for convergence of the predictor/corrector step.

```
    /* increment time */
        time+=dt;
    /* check to see if time step process has caught-up */
        if(time>=dt0)
           goto l_110;
        }
    /* convergence error in celerity */
    l_100:
```

```
    return(1);
```

The geometry for this model is based on the Clinch River, which flows out of Norris Dam and downstream 12.5 miles (20.1 km), where it enters Watts Bar Reservoir. Originally, the stream could run shallow during times of low releases from the dam, but small rock-and-cage impoundments have been added to prevent this from happening and regulate the aquatic habitat. As with the previous model, an input file is read and results written to an output file for plotting. All of the associated files can be found in the on-line archive named INFLOW.*. Typical results are:

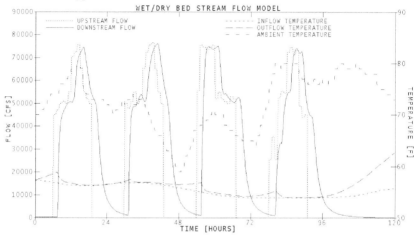

There is no sloshing in this river, as there is no downstream dam to abruptly cut off the flow and send waves back upstream.

Chapter 27. Finite Element Method

Sadly, the finite element method is more often than not presented with pomp and circumstance in a dense obscuring fog with much mathematical mumbling and a great deal of hand waving. All this is to impress the student with the wonder of it all and the breath-taking knowledge of the lecturer. I hate to break with tradition, but I'm not going to do that.

The Ensemble Hypothesis

Before we begin exploring the finite element method, we must introduce the *Ensemble Hypothesis*. In short, the ensemble hypothesis states that the whole may be greater than the sum of the individual parts. As it applies to model development, we propose that an ordered assemblage of distinguishable parts may adequately approximate the whole, in spite of the fact that no single part displays all of the characteristics of the whole.

For example, we may break an irregular domain into a collection of triangular or tetrahedral elements. We may know from experience that the distribution of the dependent variable (temperature, concentration, stress, strain, etc.) does not vary linearly over the domain, but it may over a single element. There are also boundaries distinct from the interior of the domain. We may have one type of element along the boundaries and a different type within the interior. We can put it all together in such a way that the ensemble adequately represents the whole.

Optimizing the Whole vs. the Parts

The finite difference method focuses on representing the governing equation within a single element in terms of differences that approximate the calculus. The resulting algebraic equations relate the values of the dependent variable at the nodal points, based on the differentials across each element. Various direct and indirect methods are used to solve the algebraic equations, but the approach for obtaining the equations is the same. The finite difference approximations are selected to optimally approximate the differentials. This is focusing on the *parts*.

The finite element approach is different. It starts with an unknown *basis* function that approximates the dependent variable throughout a single element. The basis function may be as simple as $T(x,y)=a+bx+cy$. The constants (a, b, and c) may be related to the temperatures at the three corners of a triangle so that the basis function matches at these three nodes. Instead of trying to minimize the error in approximating a differential within each element, we want to minimize the error in approximating the governing equation over the entire domain. We express this error as an integral over the area or volume that is the domain.

Calculus of Variations

This topic always comes up when introducing the finite element method and you may have wondered why. The reason is simple: it's the branch of mathematics in which we discuss the extrema (i.e., minima or maxima) of such things as integrals. You have no doubt seen derivations for the minimum or maximum of a function where you simply take the derivative and set it to zero. When the function is an integral over some domain, it's not so simple.

To find the optimum of an ensemble of things, you can't simply remove the integral and set what's inside to zero. It doesn't work that way and it's easy to come up with examples to illustrate this. For instance, if you want to maximize the profit from an amusement park, you must consider many things. If you raise the price of each ticket, fewer people may come. If you lower the ticket price, more people may come, but you might lose money in the process. You also can't simply charge the first rich kid who comes through a million dollars and let everyone else in free.

When it comes optimizing integrals, there are rules—new rules you may not have seen before advanced calculus, where this topic is discussed. The rules are logical and can be properly derived and also demonstrated by example. Here's the secret to the finite element method: We must find a function, such that when we integrate it over the domain and then apply the rules to arrive at the minimum error, we end up with the equation we're trying to solve in the first place. There's an expression for this process: *finding the corresponding variational problem*. Consider the following integral:

$$\iiint_{Domain} \left\{ \left(\frac{\partial T}{\partial x}\right)^2 + \left(\frac{\partial T}{\partial y}\right)^2 + \left(\frac{\partial T}{\partial y}\right)^2 \right\} dV \Rightarrow \min \qquad (27.1)$$

After we apply the rules of variational calculus to Equation 27.1, to find the minimum, we arrive at:

$$\iiint_{Domain} \left\{ \frac{\partial^2 T}{\partial x^2} + \frac{\partial^2 T}{\partial y^2} + \frac{\partial^2 T}{\partial z^2} \right\} dV = 0 \qquad (27.2)$$

Inside the integral is Laplace's equation.

$$\frac{\partial^2 T}{\partial x^2} + \frac{\partial^2 T}{\partial y^2} + \frac{\partial^2 T}{\partial z^2} = 0 \qquad (27.3)$$

If we start with a basis function that satisfies Equation 27.1 and solve it over the domain, applying the various boundary conditions, it *will* approximate Laplace's equation with the minimum residual over the entire domain. Fortunately, Laplace's equation governs a lot of problems, including: heat transfer, diffusion, magnetic and electrostatic fields, stress, and strain. Sadly, nobody has ever found the corresponding variational problem for the Navier-Stokes equations (i.e., the equations of fluid flow). Still, there are many different approximations and we will discuss some of these, but leave solutions to another book.

2D Conduction Model

We have already introduced a basis function that satisfies Equation 27.1 and that is $T(x,y)=a+bx+cy$. We will use this basis function to solve several problems governed by Laplace's equation. There are a lot of programming details necessary to read in and check a finite element model. Rather than interrupt the presentation of the finite element method at this point, these details are discussed in Appendix I. Here, we will jump into setting up and solving the equations. The nodal point equations are handled in a single concise function:

```
void NodalPointEquations()
  {
  int i,j,n1,n2,n3;
  double AKXY,A11,A12,A13,A22,A23,A33,GA3,
    X1,X12,X13,X2,X23,X3,XY,Y1,Y12,Y13,Y2,Y23,Y3;
  for(i=0;i<Ne;i++) /* element conservation equations
    */
    {
    n1=Ie[3*i]; /* node 1,2,3 from element Ie[]*/
    n2=Ie[3*i+1];
    n3=Ie[3*i+2];
    X1=Xn[n1]; /* x,y for each node from node[] */
    Y1=Yn[n1];
    X2=Xn[n2];
    Y2=Yn[n2];
    X3=Xn[n3];
    Y3=Yn[n3];
    for(j=0;j<Na[n1];j++) /* index I,J,K from Ia[] */
      Ip[Ia[Ma*n1+j]]=j;
    for(j=0;j<Na[n2];j++)
      Jp[Ia[Ma*n2+j]]=j;
    for(j=0;j<Na[n3];j++)
      Kp[Ia[Ma*n3+j]]=j;
    X12=X2-X1; /* inverse of T(x,y)=a+b*x+c*y */
    X13=X3-X1;
    X23=X3-X2;
    Y12=Y2-Y1;
    Y13=Y3-Y1;
    Y23=Y3-Y2;
    XY=X12*Y23-X23*Y12;
    AKXY=Ce[i]*Ae[i]/(XY*XY); /* k*area */
    A11= (X23*X23+Y23*Y23)*AKXY;
    A12=-(X13*X23+Y13*Y23)*AKXY;
    A13= (X12*X23+Y12*Y23)*AKXY;
    A22= (X13*X13+Y13*Y13)*AKXY;
    A23=-(X12*X13+Y12*Y13)*AKXY;
    A33= (X12*X12+Y12*Y12)*AKXY;
    Dn[n1]+=A11;
    Dn[n2]+=A22;
    Dn[n3]+=A33;
    An[Ma*n1+Ip[n2]]+=A12; /* left-hand side of A[] */
```

```
      An[Ma*n1+Ip[n3]]+=A13;
      An[Ma*n2+Jp[n1]]+=A12;
      An[Ma*n2+Jp[n3]]+=A23;
      An[Ma*n3+Kp[n1]]+=A13;
      An[Ma*n3+Kp[n2]]+=A23;
      GA3=Ge[i]*Ae[i]/3; /* generation term */
      Bn[n1]+=GA3;  /* right-hand side of A[]X[]=B[] */
      Bn[n2]+=GA3;
      Bn[n3]+=GA3;
      }
  }
```

The core of this section of code is the inverse of the following matrix:

$$\begin{bmatrix} 1 & x_1 & y_1 \\ 1 & x_2 & y_2 \\ 1 & x_3 & y_3 \end{bmatrix} \qquad (27.4)$$

There are three types of boundary conditions: isothermal ($T=...$), heat flux ($q=...$), and convection ($q=h*(T-To)$). These are added in the following simple function:

```
void BoundaryConditions()
  {
  int i,j,j1,j2,k,n,n1,n2;
  double b1,b2,hs,hs3,hs6,hst2,q,s;
  for(n=0;n<Nb;n++)
    {
    i=Ib[n];
    j1=Jb[n];
    j2=(j1+1)%3;
    n1=Ie[3*i+j1];
    n2=Ie[3*i+j2];
    k=Kb[n];
    b1=B1[n];
    b2=B2[n];
    s=Se[3*i+j1];
    if(k==0) /* isothermal */
      {
      Dn[n1]=1;
      Bn[n1]=b1;
      memset(An+Ma*n1,0,Ma*sizeof(double));
      Dn[n2]=1;
      Bn[n2]=b1;
      memset(An+Ma*n2,0,Ma*sizeof(double));
      }
    else if(k==1) /* heat flux */
      {
      q=b1*s/2;
      Bn[n1]+=q;
```

```
        Bn[n2]+=q;
      }
    else /* convection */
      {
      for(j=0;j<Na[n1];j++)
         Ip[Ia[Ma*n1+j]]=j;
      for(j=0;j<Na[n2];j++)
         Jp[Ia[Ma*n2+j]]=j;
      hs=b1*s;
      hs3=hs/3;
      hs6=hs/6;
      hst2=hs*b2/2;
      Bn[n1]+=hst2;
      Bn[n2]+=hst2;
      Dn[n1]+=hs3;
      Dn[n2]+=hs3;
      An[Ma*n1+Ip[n2]]+=hs6;
      An[Ma*n2+Jp[n1]]+=hs6;
      }
    }
  }
```

That's it. The rest is solving a sparse matrix. There are several methods available to accomplish this. The method used in this case is Successive Over Relaxation (SOR). This and several other methods are discussed in Appendix J. All of the files associated with this example can be found in the on-line archive in a folder named examples\finite elements\2D-conduction named con2d.*. The following is typical input:

```
LOWER HALF OF A SYMMETRIC CONVECTING FIN IN AN ISO-
    THERMAL SUBSTRATA
NE NN NB IPRT MAP NITER   RELAX   DTMAX
65 49 31  1   1    100     1.5    0.1
ELEMENT SPECIFICATIONS
1  1  2   3 0 .2
2  2  4   3 0 .2
3  3 44   5 0 .2
4  4  6  44 0 .2
etc...
BOUNDARY SPECIFICATIONS
1  1 1 2 0 0
2  2 1 2 0 0
3  4 1 2 0 0
4  6 1 2 0 0
etc...
NODAL POINT LOCATIONS
1 .00 .08
2 .02 .08
3 .02 .07
4 .04 .08
```

In order to adequately visualize the results, an advanced graphics program (such as TP2) is necessary. The program (CON2D) creates all of the necessary output files so that TP2 can display the results. Three examples are provided: fin, star, and tree. The fin below is what you might find on the head or cylinder of a motorcycle engine, dissipating heat to the air passing over it.

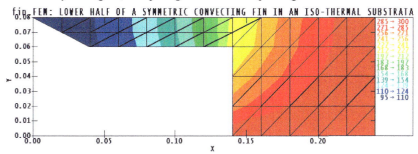

This next problem has internal heat generation at the center of the red zone:

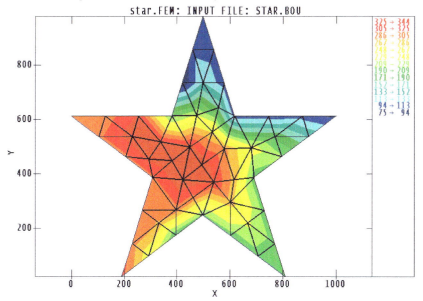

This third problem is just a three-pronged fin. There is a batch file to solve and plot each one of these examples (e.g., _solve_tree.bat). This is the results for the third problem:

213

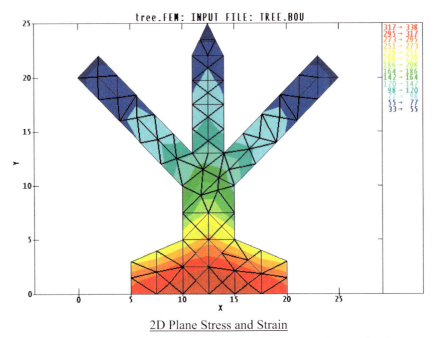

2D Plane Stress and Strain

The next problem we will consider is two-dimensional (x,y plane) stress and strain. The following figure illustrates plane stress:

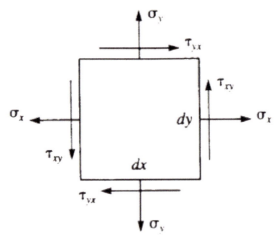

In two dimensions (x,y) the stress tensor (vector) has three parts:

$$[\sigma_x \quad \sigma_y \quad \tau_{xy}]^T \tag{27.5}$$

Plane strains are illustrated in this next figure:

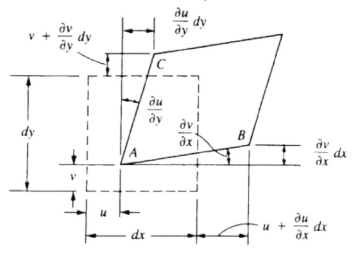

In two dimensions (x,y) the strain tensor (vector) also has three parts:

$$\left[\varepsilon_x = \frac{\partial u}{\partial x} \quad \varepsilon_y = \frac{\partial v}{\partial y} \quad \gamma_{xy} = \frac{\partial u}{\partial y} + \frac{\partial v}{\partial x}\right]^T \quad (27.6)$$

The stress and strain are related by Hooke's Law, which can be expressed in vector form:

$$\begin{Bmatrix} \sigma_x \\ \sigma_y \\ \tau_{xy} \end{Bmatrix} = [D] \begin{Bmatrix} \varepsilon_x \\ \varepsilon_y \\ \gamma_{xy} \end{Bmatrix} \quad (27.7)$$

For plane stress the matrix [D] is given by:

$$[D] = \frac{E}{1-v^2} \begin{bmatrix} 1 & v & 0 \\ v & 1 & 0 \\ 0 & 0 & \frac{1-v}{2} \end{bmatrix} \quad (27.8)$$

where E is Young's modulus and v is Poisson's Ratio.

For plane strain the matrix [D] is given by:

$$[D] = \frac{E}{(1+v)(1-2v)} \begin{bmatrix} 1-v & v & 0 \\ v & 1-v & 0 \\ 0 & 0 & \frac{1}{2}-v \end{bmatrix} \quad (27.9)$$

Because the stress and strain satisfy Laplace's equation, we use the same basis function, $f(x,y)=a+bx+cy$. In this section, we will only consider triangular elements, as you can build whatever you want out of them and there are various programs available to discretize most any shape. The area of a triangle is given by the determinant of the following 3x3 matrix:

$$A = \frac{1}{2} \begin{bmatrix} 1 & x_1 & y_1 \\ 1 & x_2 & y_2 \\ 1 & x_3 & y_3 \end{bmatrix} \quad (27.10)$$

Next, we define a basis vector containing the basis function at the three nodes of a triangular element:

$$[N] = \frac{1}{2A} \begin{bmatrix} a_1 + b_1 x + c_1 y \\ a_2 + b_2 x + c_2 y \\ a_3 + b_3 x + c_3 y \end{bmatrix} \quad (27.11)$$

Following Equation 27.2, we take the partial derivatives of the basis vector and integrate over the volume of an element to arrive at the following, known as the stiffness matrix, [K]:

$$[K] = V[B]^T[D][B] \quad (27.12)$$

where V is the volume, equal to the area, A, times the thickness, t, and [B] is the matrix resulting from the differentiation of [N]:

$$[B] = \frac{1}{2A} \begin{bmatrix} y_{23} & 0 & y_{31} & 0 & y_{12} & 0 \\ 0 & x_{32} & 0 & x_{13} & 0 & x_{21} \\ x_{32} & y_{23} & x_{13} & y_{31} & x_{21} & y_{12} \end{bmatrix} \quad (27.13)$$

where $x_{ij}=x_i-x_j$ and $y_{ij}=y_i-y_j$. These equations are simple. Most of the effort implementing the method is bookkeeping. All of the files associated with this section can be found in the on-line archive in the folder examples\finite element\plane strain. We begin with a typical input file:

```
8 6 1 2 nodes elements pointloads bearings
0 0 x y
100 0
```

```
100 100
0 100
200 0
200 100
300 0
300 100
1 2 3 200000 0.3 0.5   i j k Young Poisson thickness
1 3 4 200000 0.3 0.5
2 5 6 200000 0.3 0.5
2 6 3 200000 0.3 0.5
5 7 8 200000 0.3 0.5
5 8 6 200000 0.3 0.5
8 0 -10000   n Fx Fy
1 1 1        n Rx Ry
4 1 0
```

The number of nodes, elements, point loads, and bearings (constraints) are on the first line. After that we have x,y for each node on sequential lines until all are specified. Following that, we have the three nodes that define each element, along with Young's modulus, Poisson's Ratio, and the thickness for each element. This allows us to handle composites in which the properties vary from element-to-element. It doesn't matter what the units are, as long as they are consistent (length, force, force/length, and force/length^2).

After the nodes and elements, the point loads (force per unit thickness). These are specified by the number of the node to which the force is applied, along with the x and y components. The bearings (horizontal and/or lateral constraints) are defined in similar manner: node, x constraint, y constraint. A value of 0 indicates the point is not constrained and a value of 1 indicates that it is constrained.

This code (psfem.c) wants the elements to be defined in counter-clockwise orientation. To facilitate input, we perform a simple test. If the area of the element is negative, we swap the order of the elements. That way, it doesn't matter how you order the nodes in the input file.

```
double Area(int i,int j,int k)
  {
  double x1,x2,x3,y1,y2,y3;
  x1=Node[i].x;
  x2=Node[j].x;
  x3=Node[k].x;
  y1=Node[i].y;
  y2=Node[j].y;
  y3=Node[k].y;
  return(((x2-x1)*(y3-y2)-(x3-x2)*(y2-y1))/2.);
  }
    if(Area(i,j,k)>0.)
      {
      Elem[l].i=i;
      Elem[l].j=j;
```

```
      Elem[l].k=k;
      }
    else
      {
      Elem[l].i=i;
      Elem[l].j=k;
      Elem[l].k=j;
      }
    Elem[l].ym=ym;
    Elem[l].nu=nu;
    Elem[l].tk=tk;
    }
```

After reading the input file, we assemble the global matrix:

```
  for(i=0;i<elems;i++)
    {
    stif=Stiff(Node[Elem[i].i].x,Node[Elem[i].i].y,
               Node[Elem[i].j].x,Node[Elem[i].j].y,
               Node[Elem[i].k].x,Node[Elem[i].k].y);
    getKe(stif,Elem[i].ym,Elem[i].nu,Elem[i].tk,ke);
    if(!Test(ke))
       Abort(__LINE__,"element %i is degenerate",i+1);
    Global(ke,i);
    }
```

Within this loop, we calculate the stiffness matrix [K] for each element:

```
STIF Stiff(double x1,double y1,double x2,double
   y2,double x3,double y3)
   {
   static STIF stif;
   stif.x1=x1;
   stif.y1=y1;
   stif.x2=x2;
   stif.y2=y2;
   stif.x3=x3;
   stif.y3=y3;
   stif.x13=x1-x3;
   stif.x32=x3-x2;
   stif.x21=x2-x1;
   stif.x23=x2-x3;
   stif.y13=y1-y3;
   stif.y23=y2-y3;
   stif.y31=y3-y1;
   stif.y12=y1-y2;
   return(stif);
   }
void getB(STIF stif,double*B)
   {
   double det;
   det=stif.x13*stif.y23-stif.x23*stif.y13;
   B[ 0]=stif.y23/det;
```

```
    B[ 1]=0.;
    B[ 2]=stif.y31/det;
    B[ 3]=0.;
    B[ 4]=stif.y12/det;
    B[ 5]=0.;
    B[ 6]=0.;
    B[ 7]=stif.x32/det;
    B[ 8]=0.;
    B[ 9]=stif.x13/det;
    B[10]=0.;
    B[11]=stif.x21/det;
    B[12]=stif.x32/det;
    B[13]=stif.y23/det;
    B[14]=stif.x13/det;
    B[15]=stif.y31/det;
    B[16]=stif.x21/det;
    B[17]=stif.y12/det;
    }

void getD(STIF stif,double ym,double nu,double*D)
    {
    double c;
#if(defined(PlaneStress))
    c=ym/(1.0-nu*nu);
    D[0]=c;
    D[1]=nu*c;
    D[2]=0.;
    D[3]=nu*c;
    D[4]=c;
    D[5]=0.;
    D[6]=0.;
    D[7]=0.;
    D[8]=(1.-nu)*0.5*c;
#elif(defined(PlaneStrain))
    c=ym/((1.+nu)*(1.-2.*nu));
    D[0]=(1.-nu)*c;
    D[1]=nu*c;
    D[2]=0.;
    D[3]=nu*c;
    D[4]=(1.-nu)*c;
    D[5]=0.;
    D[6]=0.;
    D[7]=0.;
    D[8]=(0.5-nu)*c;
#else
    #error neither PlaneStrain nor PlaneStress
#endif
    }
```

```
void getKe(STIF stif,double ym,double nu,double
  tk,double*ke)
{
int i,j;
double at,B[3*6],BT[6*3],BTD[6*3],D[3*3];
getB(stif,B);
Transpose(B,BT,3,6);
getD(stif,ym,nu,D);
Multiply(BT,D,BTD,6,3,3);
Multiply(BTD,B,ke,6,3,6);
at=tk*(stif.x13*stif.y23-stif.x23*stif.y13)/2.;
for(i=0;i<6;i++)
  for(j=0;j<6;j++)
    ke[6*i+j]*=at;
}
```

We build the global matrix by inserting the stiffness matrix for each element in the proper place for the nodes that form the element:

```
void Global(double*ke,int e)
{
int i,j,k,l,m[6];
m[0]=(Elem[e].i)*DOF;
m[1]=(Elem[e].i)*DOF+1;
m[2]=(Elem[e].j)*DOF;
m[3]=(Elem[e].j)*DOF+1;
m[4]=(Elem[e].k)*DOF;
m[5]=(Elem[e].k)*DOF+1;
for(i=0;i<6;i++)
  {
  k=m[i];
  for(j=0;j<6;j++)
    {
    l=m[j]-k;
    if(l>=0)
      globe(k,l)+=ke[6*i+j];
    }
  }
}
```

We add the point loads as forces on the respective nodes in the solution vector:

```
if(loads)
  {
  printf("  applying poing loads\n");
  for(i=0;i<loads;i++)
    {
    solu[(Load[i].n-1)*DOF]+=Load[i].Fx;
    solu[(Load[i].n-1)*DOF+1]+=Load[i].Fy;
    }
  }
```

We add the bearings (horizontal and/or vertical constraints) using the penalty method. This is a round-about way of saying that we connect a really stiff spring between the node and its original location.

```
if(bears)
  {
  printf("  applying bearings using penalty
  method\n");
  pen=globe(1,0);
  for(i=1;i<nvar;i++)
    pen=max(pen,globe(i,0));
  pen=log10(pen);
  pen=pow(10.,ceil(pen)+10.);
  for(i=0;i<bears;i++)
    {
    if(Bear[i].Rx)
      globe((Bear[i].n-1)*DOF,0)*=pen;
    if(Bear[i].Ry)
      globe((Bear[i].n-1)*DOF+1,0)*=pen;
    }
  }
```

We then solve the system of linear equations using Gauss Elimination. We must always use row pivoting with this method, but we can't use column pivoting in this case, because the matrix is sparse and we don't ever deal with the full matrix.

```
void Gauss(double*x)
  {
  int i,ii,j,k,l,m,n;
  double aik,akj,piv,sum;
  for(ii=0;ii<nvar-1;ii++)
    {
    piv=globe(ii,0);
    m=ii+span;
    if(m>nvar)
      m=nvar;
    for(i=ii+1;i<m;i++)
      {
      aik=globe(ii,i-ii);
      for(l=0;l<span-1;l++)
        {
        if(i-ii+l<span)
          {
          akj=globe(ii,i-ii+l);
          globe(i,l)-=aik/piv*akj;
          }
        }
      x[i]=x[i]-aik/piv*x[ii];
      }
    }
```

```
piv=globe(nvar-1,0);
x[nvar-1]=x[nvar-1]/piv;
for(k=nvar-2;k>=0;k--)
  {
  sum=0.;
  n=span-1;
  if(n+k>=nvar)
     n=nvar-k-1;
  for(j=1;j<=n;j++)
     sum=sum+globe(k,j)*x[j+k];
  x[k]=1./globe(k,0)*(x[k]-sum);
  }
}
```

After solving for the unknown stresses, we calculate the strains by applying the stiffness matrix to each element:

```
for(i=0;i<elems;i++)
  {
  stif=Stiff(Node[Elem[i].i].x,Node[Elem[i].i].y,
             Node[Elem[i].j].x,Node[Elem[i].j].y,
             Node[Elem[i].k].x,Node[Elem[i].k].y);
  getB(stif,B);
  k[0]=Elem[i].i*DOF;
  k[1]=Elem[i].i*DOF+1;
  k[2]=Elem[i].j*DOF;
  k[3]=Elem[i].j*DOF+1;
  k[4]=Elem[i].k*DOF;
  k[5]=Elem[i].k*DOF+1;
  for(j=0;j<6;j++)
     disp[j]=defo[k[j]];
  Multiply(B,disp,Elem[i].st,3,6,1);
  getD(stif,Elem[i].ym,Elem[i].nu,D);
  Multiply(D,Elem[i].st,Elem[i].ss,3,3,1);
  }
```

That's all there is to it. There's an optional zoom factor to exaggerate the nodal point displacements for visualization. It defaults to 5, but you can specify it as a second argument when you launch the program. The first argument is the input file name. The simplest problem we will consider is the one prescribed in the sample input file above (canti.fem). This is a cantilever beam, rigidly attached on the left with a downward force on the right end. The resulting nodal point displacements are shown in this next figure:

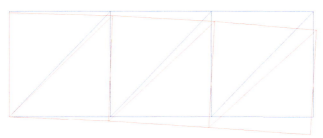

The blue lines are the undeformed elements and the red lines are the deformed ones, with an exaggeration factor (zoom) of 5. Output is written to two files. The first (psfem.out) contains the stresses and strains and the second (psfem.2dv) contains two *bodies* (a group of nodes and elements). The first is the undeformed and the second is the deformed with exaggeration. The next example is a simply-loaded beam (beam1.fem), supported at the ends with a downward force in the middle:

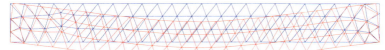

The next example is a much more finely discretized beam (beam2.fem) that is clamped at both ends with a load in the middle.

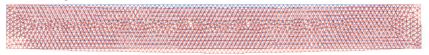

The next example is a flange with a hole in it that is stretched (flange.fem):

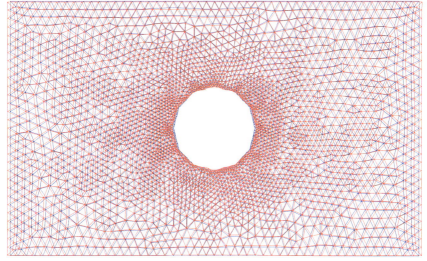

The last example is a loaded hook (hook.fem):

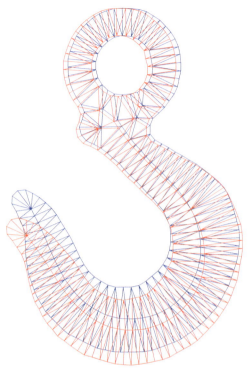

This is what the output of the program looks like. From this you can see the order of operations from reading the input file to summarizing the results:

```
Plane Strain Finite Element Model
reading input: hook.fem
  reading 238 nodes
  reading 298 elements
  reading 2 loads
  reading 2 bearings
begin solution
  sizing matrices
  applying point loads
  applying bearings using penalty method
  solving matrix for results
  calculating displacements
saving results: psfem.out
creating 2D view file: psfem.2dv
```

Chapter 28. Boundary Element Method

The boundary element method arises from Green's Lemma[42]. This remarkable relationship between the integral of a function over an area and the integral of a corresponding function around the perimeter of the same domain is usually covered in advanced calculus. Green's Lemma can be expressed by the following integral:

$$\iint \nabla^2 \varphi \, dA = \int \frac{\partial \varphi}{\partial n} \, dS \qquad (28.1)$$

In Equation 7.1 φ is the potential, dA is the differential area within the domain, dS is a differential distance along the boundary, and n is the normal (perpendicular) at each location along the boundary. The potential could be simply that (i.e., electrostatic potential or invicid flow), temperature, concentration, stress, strain, or anything else that satisfies Laplace's equation.

We must also have a *fundamental* solution to Laplace's equation—in this case, a general, homogeneous solution (i.e., works for any case and is zero on the right-hand side). The fundamental solution is similar to the basis function postulated in the finite element method. Instead of satisfying the variational statement, this function must satisfy the equation itself. The derivation is a bit circuitous, except in polar coordinates.

$$\varphi = \ln\left(\frac{1}{r}\right) \qquad (28.2)$$

We can at least show by substitution that Equation 7.2 satisfies Laplace's equation in polar coordinates:

$$\frac{1}{r}\frac{\partial}{\partial r}\left(r\frac{\partial \varphi}{\partial r}\right) + \frac{1}{r^2}\frac{\partial^2 \varphi}{\partial \theta^2} = 0 \qquad (28.3)$$

For Cartesian (x,y) coordinates, r in the above equation is the distance from some as yet unspecified location $(x=a, y=b)$, or $r^2=(x-a)^2+(y-b)^2$. We further propose that the potential, φ, and the derivative with respect to the normal, $\partial \varphi / \partial n$, have some specific, finite, non-trivial value at each of the points along the boundary. Consider two points along the boundary (1 and 2). We can write Equation 7.2 for these two points:

---

[42] George Green (1729-1841): British mathematical physicist best known for work with electric fields and magnetism.

$$\varphi(r) = \left\{ \frac{\varphi_1(r-r_2)}{\ln\left(\frac{1}{r_1}\right)} - \frac{\varphi_2(r-r_1)}{\ln\left(\frac{1}{r_2}\right)} \right\} \frac{\ln\left(\frac{1}{r}\right)}{(r_1-r_2)} \qquad (28.4)$$

At x_1, y_1 $r=r_1$ and at x_2, y_2 $r=r_2$ so that $\varphi(r_1) = \varphi_1$ and $\varphi(r_2) = \varphi_2$; therefore, Equation 7.4 is the particular form of the fundamental equation that satisfies Laplace's equation and matches at these two points along the boundary. We can construct a similar equation for the right side of Equation 7.1. When integrate Equation 7.4 from point 1 to point 2, we will get some constant times φ_1 plus some other constant times φ_2. We will use $H_{i,j}$ to represent these constants on the left side and $G_{i,j}$ to represent the corresponding constants on the right side of Equation 7.1. The indices i and j indicate each segment along the boundary and each pair corresponding to each segment. The resulting set of equations can be written:

$$\sum_{i=}^{n}\sum_{j=1}^{n} H_{i,j}\varphi_j = \sum_{i=}^{n}\sum_{j=1}^{n} G_{i,j} \frac{\partial \varphi_j}{\partial n} \qquad (28.5)$$

Equation 7.5 constitutes a set of simultaneous linear equations for the potential function and its derivative at the n points along the boundary. We will need three different formulas for the integrals. We can readily integrate this equation around the boundary except at the two points (here, 1 and 2). At those points $r=r_1$ or $r=r_2$ and the standard result is indeterminate. We use a different formula for that one segment. We use these two formulas for points along the boundary. For points inside the boundary we use a third formula. We need this third integral to evaluate the results of the solution inside the boundary.

I have been vague up until this point on exactly what formulas are integrated and how, sparing you the gory details. In most applications, numerical integration (e.g., Gauss Quadrature) is used, because the analytical integral is unknown to the programmer. In fact, that's the way I present this as an example in my book *Numerical Calculus*. Here, you have the benefit of the analytical solution, which is precise, instead of a numerical solution that is approximate. Of course, I didn't figure this out the hard way. I used Maple® to do that for me. The resulting formulas are indeed tedious. The functions that accomplish these three integrals (adjacent, opposite, and internal) are:

```
void AdjacentBoundary(double X1,double Y1,double
   X2,double Y2,double*G1,double*G2)
{
double A,dS;
dS=hypot(X2-X1,Y2-Y1);
A=log(dS);
*G1=dS*(1.5-A)/2;
```

```
  *G2=dS*(0.5-A)/2;
  }
void OppositeBoundary(double Xp,double Yp,double
   X1,double Y1,double X2,double
   Y2,double*G1,double*G2,double*H1,double*H2)
  {
  double pX=Xp-X1;
  double pY=Yp-Y1;
  double dX=X2-X1;
  double dY=Y2-Y1;
  double SS=dX*dX+dY*dY;
  double dS=sqrt(SS);
  double t3=-pX*dY+dX*pY;
  double t5=1/SS;
  double t6=t3*t5;
  double t7=pX*pX;
  double t8=pY*pY;
  double t10=log(t7+t8);
  double t12=-t3*t5;
  double t13=dX*dX;
  double t14=dX*pX;
  double t16=dY*dY;
  double t17=dY*pY;
  double t20=log(t13-2*t14+t7+t16-2*t17+t8);
  double t22=t13+t16-t17-t14;
  double t24=-1/t3;
  double t26=atan(-t22*t24);
  double t29=t17+t14;
  double t31=atan(t29*t24);
  double t42=1/SS/dS;
  double t43=t13*dX;
  double t45=2*t43*pX;
  double t46=t7*t13;
  double t47=t8*t16;
  double t48=t7*t16;
  double t49=t8*t13;
  double t51=4*t14*t17;
  double t52=t16*dY;
  double t54=2*t52*pY;
  double t57=2*t13*dY*pY;
  double t58=t16*dX;
  double t60=2*t58*pX;
  double t64=t13*t13;
  double t65=t13*t16;
  double t66=2*t65;
  double t67=t16*t16;
  double t68=t64+t46+t47-t48-t54+t66-t45+t67+t51-t57-
    t60-t49;
  double t74=dY*t8*dX;
  double t78=pX*pY*t13;
```

```
    double t80=t7*dX*dY;
    double t82=pX*t16*pY;
    double t84=-t52*pX+t43*pY-t74-t13*pX*dY-
      t78+t80+t82+t58*pY;
    double t101=t80+t82-t78-t74;
    *H1=-t6*t10/2-t12*t20/2-t22*t5*t26+t22*t5*t31;
    *H2=-t12*t10/2-t6*t20/2-t29*t5*t26+t29*t5*t31;
    *G1=-t42*(t45-t46-t47+t48+t49-t51+t54+t57+t60)*t10/4-
      t42*t68*t20/4-t42*t84*t26+t42*t84*t31-t42*(-6*t65-
      3*t64+t57+t60-3*t67+t45+t54)/4;
    *G2=t42*(t48-t46-t51-t47+t49)*t10/4+t42*(-t64-t67-
      t48+t47+t51+t46-t49-t66)*t20/4+t42*t101*t26-
      t42*t101*t31+t42*(t57+t60+t66+t45+t54+t64+t67)/4;
    }
  int InternalIntegral(double Xp,double Yp,double
    X1,double Y1,double X2,double
    Y2,double*G1,double*G1x,double*G1y,double*G2,double*
    G2x,double*G2y,double*H1,double*H1x,double*H1y,doubl
    e*H2,double*H2x,double*H2y)
    {
    double pX=Xp-X1;
    double pY=Yp-Y1;
    double dX=X2-X1;
    double dY=Y2-Y1;
    double SS=dX*dX+dY*dY;
    double dS=sqrt(SS);
    double t3=1/SS/dS;
    double t4=dY*dY;
    double t5=t4*t4;
    double t6=pY*pY;
    double t7=dX*dX;
    double t8=t6*t7;
    double t9=pX*pX;
    double t10=t9*t4;
    double t11=t4*dY;
    double t13=2*t11*pY;
    double t14=t7*dX;
    double t16=2*t14*pX;
    double t17=t7*t4;
    double t18=2*t17;
    double t19=t9*t7;
    double t20=t6*t4;
    double t21=t7*t7;
    double t22=pY*t7;
    double t24=2*t22*dY;
    double t25=t4*pX;
    double t27=2*t25*dX;
    double t28=pX*dX;
    double t29=pY*dY;
    double t31=4*t28*t29;
```

```
double t32=t5-t8-t10-t13-t16+t18+t19+t20+t21-t24-
  t27+t31;
double t36=t7-2*t28+t9+t4-2*t29+t6;
double t37=log(t36);
double t41=t6+t9;
double t42=log(t41);
double t46=pY*dX;
double t48=dX*t9;
double t49=t48*dY;
double t50=pY*t4;
double t51=t50*pX;
double t52=t22*pX;
double t53=t7*pX;
double t55=dX*dY;
double t56=t55*t6;
double t58=4*t3*(t14*pY-pX*t11+t46*t4+t49+t51-t52-
  t53*dY-t56);
double t59=t7+t4-t28-t29;
double t61=t46-pX*dY;
double t62=1/t61;
double t64=atan(t59*t62);
double t66=t28+t29;
double t68=atan(t66*t62);
double t76=t55*pY;
double t77=2*t76;
double t78=t4*dX;
double t79=-t14+t53+t77-t78-t25;
double t84=2*t22;
double t85=t7*dY;
double t86=2*t85;
double t87=t28*dY;
double t88=4*t87;
double t89=2*t50;
double t90=2*t11;
double t92=t3*(-t84-t86+t88+t89-t90);
double t95=t14+t78;
double t98=2*t87;
double t99=-t22-t85+t98+t50-t11;
double t104=2*t14;
double t105=2*t53;
double t106=2*t78;
double t107=2*t25;
double t108=4*t76;
double t110=t3*(t104-t105+t106+t107-t108);
double t114=2*t3*(t85+t11);
double t123=4*t3*(t52+t56-t49-t51);
double t129=-t53-t77+t25;
double t135=t3*(t84-t88-t89);
double t140=-t22+t98+t50;
double t146=t3*(-t105-t108+t107);
```

```
double t150=1/SS;
double t151=-t61*t150;
double t153=t61*t150;
double t155=2*t59*t150;
double t159=dY*t6;
double t160=dY*t9;
double t161=t159+t160;
double t162=1/t41;
double t164=t150*t37;
double t167=t150*t42;
double t169=dX*t6;
double t170=-t169-t48;
double t171=2*t170*t162;
double t172=t150*t64;
double t174=t150*t68;
double t184=2*t161*t162;
double t193=2*t66*t150;
double t197=-t159-t160-t11-t85+t89+t98;
double t198=1/t36;
double t207=(t104+2*t169+2*t48+t106-4*t53-t108)*t198;
double t214=t14-t105+t169-t77+t78+t48;
double t225=(-t88-4*t50+t90+2*t160+t86+2*t159)*t150;
*G1=-t3*t32*t37/4-t3*(t13+t16+t10+t8-t19-t20-
  t31+t27+t24)*t42/4-t58*t64/4-t58*t68/4-t3*(t16+t13-
  3*t5+t24+t27-6*t17-3*t21)/4;
*G1x=-(t3*t79*t37/2-
  t3*t79*t42/2+t92*t64/2+t92*t68/2+t3*t95);
*G1y=-(t3*t99*t37/2-
  t3*t99*t42/2+t110*t64/2+t110*t68/2+t114/2);
*G2=-t3*(t18-t19+t8-t20+t10-t31+t5+t21)*t37/4-t3*(-
  t8+t19+t31+t20-t10)*t42/4-t123*t64/4-t123*t68/4-
  t3*(-t16-t13-t18-t27-t5-t21-t24)/4;
*G2x=-(t3*t129*t37/2-
  t3*t129*t42/2+t135*t64/2+t135*t68/2-t3*t95);
*G2y=-(-t3*t140*t37/2+t3*t140*t42/2-t146*t64/2-
  t146*t68/2-t114/2);
*H1=-t151*t37/2-t153*t42/2-t155*t64/2-t155*t68/2;
*H1x=-(t161*t162*t164/2-
  t161*t162*t167/2+t171*t172/2+t171*t174/2+(t22+t50)*t
  162*t150);
*H1y=-(t170*t162*t164/2-t170*t162*t167/2-t184*t172/2-
  t184*t174/2-(t25+t53)*t162*t150);
*H2=-t153*t37/2-t151*t42/2-t193*t64/2-t193*t68/2;
*H2x=-(t197*t198*t164/2-
  t197*t198*t167/2+t207*t172/2+t207*t174/2+(t11+t85-
  t22-t50)*t198*t150);
*H2y=-(t214*t150*t198*t37/2-
  t214*t150*t198*t42/2+t225*t198*t64/2+t225*t198*t68/2
  +(t53+t25-t14-t78)*t150*t198);
return(0);
```

}

You read in the boundary points and the boundary conditions. You build the equations using the preceding integrals. Then you solve the simultaneous equations for *2n* unknowns (the potential and its derivative at each of the *n* points) and that's it! You use the third integral to find the value of the potential and its derivative at any point inside the boundary. You can then use some graphics program (e.g., TP2) to plot lines of constant potential throughout the domain, now that you know the value at however many points are desired. You can also calculate velocity vectors (or heat flux vectors or magnetic field vectors) from the potential and it's normal derivative that you got from the internal integral function.

You will find the program source code and 8 examples in the on-line archive in the folder examples\boundary element. The following is a typical input file (curl.bem):

```
Ye & McCorquodale's 270° bend
152 -5000 number of boundary nodes, number of internal
    calculation points
 0.00   0.00 0 0
 0.43   0.00 0 0
 0.86   0.00 0 0
 1.30   0.00 0 0
 1.73   0.00 0 0
 2.16   0.00 0 0
 2.59   0.00 0 0
 3.03   0.00 0 0
```

The output for this example is:

```
PFLOW/2.31: Boundary Element Method for Potential
    Fields
  by Dudley J. Benton, dudley.benton@gmail.com
input file: curl.bem
title: Ye & McCorquodale's 270° bend
number of boundary points .......... 152
number of internal points .......... -5000
allocating memory .................. 372 KB
reading boundary points ............ OK
total number of boundaries ......... 1
number of internal regions ......... 0
number of essential BCs ............ 152
number of normal BCs ............... 0
number of tangent BCs .............. 0
sorting boundaries ................. OK
essential boundary conditions ...... 152
number of simultaneous equations ... 152
allocating memory .................. 556 KB
nodal point equations .............. 152
resorting boundary conditions ...... OK
simultaneous equations ............. 152
```

```
reordering solution set ............ OK
calculating tangentials ............ OK
results output file ................ PFLOW.OUT
boundary output file ............... PFLOW.P2D
velocity output file ............... PFLOW.V2D
potential file ..................... PFLOW.TRI
plot file .......................... PFL0W.TP2
plot command file .................. PFLOW.TP2
for velocity vectors run TP2 PFLOW.TP2
for potential contours run TP2 PFL0W.TP2
```

The potential field inside the domain is:

The velocity vectors are:

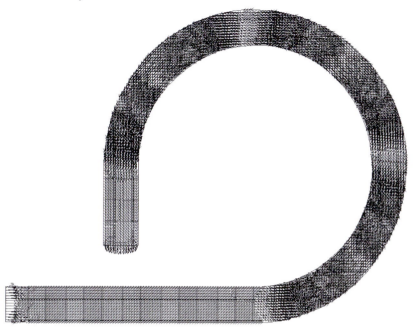

There's also a lake with two inlets, two outlets, and two islands (lake.bem). Potential is:

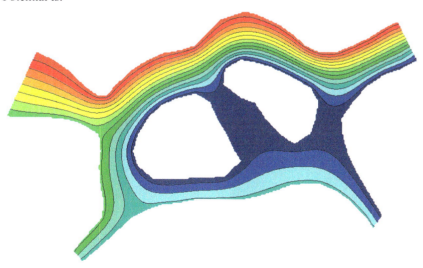

The velocity vectors are:

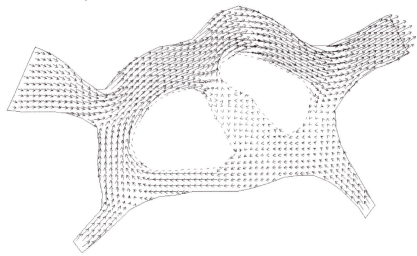

There is also a natural draft cooling tower in a slight cross wind (recirc.bem). The potential is:

The velocity vectors are:

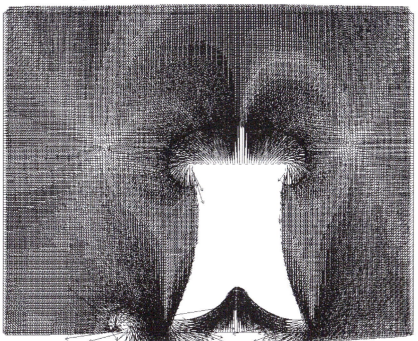

Here's a graph of flow through a coal-fired boiler (boiler.bem):

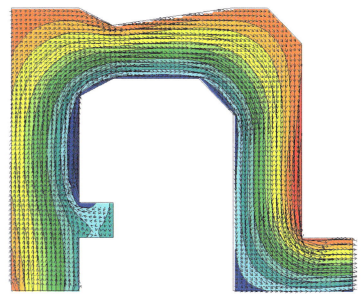

This next figure shows flow through a fork on the Holston River (reservior.bem):

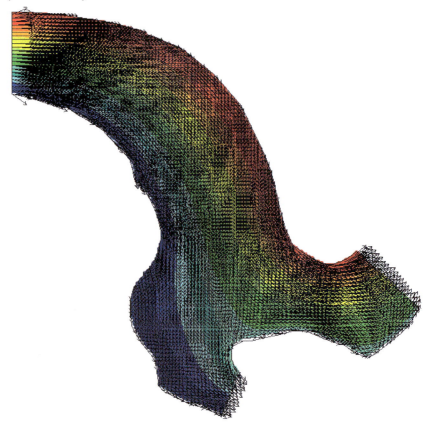

The picture on the cover shows flow through Tarpon Bend (bend.bem). You can use this method to calculate the flow through any shape.

All of the files to create the above can be found in the on-line archive in folder examples\boundary element\NTRC.

Appendix A: Steam Properties - A Practical Example

It's easy to provide trivial examples tailored to accentuate the strengths of one method over another, thus rigging the outcome. It is quite another thing to provide meaningful applications. We will now consider one such example: thermodynamic properties of steam. Properties are calculated many times in a spreadsheet or model. Each calculation involves a length function call with nested loops and transcendental functions, thus it can take a significant amount of time, even on modern computers.

No one has used Keenan, Keyes, Hill, and Moore's (KKHM) 1969 steam properties[43] for decades. They were never widely used, even when they first came out. They were a little more accurate than the 1967 properties[44], which had been endorsed by the ASME and still used by the General Electric Steam Turbine Division to this day (ASME67). There are three more sets of steam properties you should know about: the 1984, published by the NBS/NRC[45] (the National Bureau of Standards is now called the National Institute of Standards and Testing), the IAPWS-SF95[46], and the IAPWS-IF97.[47]

The KKHM properties are based on an elegant formulation and for that reason alone are of significant historical value. The NBS/NRC took this one step farther. It is a great shame that these two works have fallen into disuse. The IAPWS-SF95 is also an elegant formulation, though this work would not have been possible without the trailblazing authors of KKHM and NBS/NRC. The ASME67 and IAPWS-IF97 are both mathematical atrocities and should have been tossed in the recycle bin long ago.

Why are the steam properties so computationally intensive? Because we most often know temperature and pressure, but these are not always independent variables and, therefore, can't adequately represent the behavior of a substance over a wide range of conditions. All chemically-stable exhibit saturation stated in which the liquid or solid and vapor are in equilibrium. These same substances also exhibit two very interesting points: 1) the critical

---

[43] Keenan, J. H., Keyes, F. G., Hill, P. G., and Moore, J. G., *Steam Tables*, John Wiley & Sons, Inc., 1969.
[44] Meyer, C. A., McClintock, R. B., Silvestri, G. J., and Spencer, R. C., Jr., *Thermodynamic and Transport Properties of Steam*, American Society of Mechanical Engineers, 1967.
[45] Haar, L., Gallagher, J. S., and Kell, G. S., *Steam Tables*, NBS/NRC printed by Hemisphere, distributed by McGraw-Hill, 1984.
[46] Friend, D. G. and Dooley, R. B., *Revised Formulation for the Thermodynamic Properties of Ordinary Water Substance for General and Scientific Use*, The International Association for the Properties of Water and Steam, 1995
[47] Research and Technology Committee on Water and Steam in Thermal Power Systems, *ASME Steam Properties for Industrial Use*, The American Society of Mechanical Engineers.

point, at which the liquid and vapor are indistinguishable, and 2) the triple point, at which the solid, liquid, and vapor coexist in equilibrium.

Temperature and density are always independent for such substances and the properties are continuous in these two variables. Therefore, the formulation must also be continuous in these two variables. This is why the ASME67 and IAPWS-IF97 are mathematical atrocities—they aren't continuous and are jumble of brute-force curve fitting. Keenan, Keyes, Hill, and Moore weren't the first to publish it, but they were the first to implement this important observation: all of the thermodynamic properties of a substance can be expressed in terms of the Helmholtz Free Energy (HFE) or its derivatives. Find the HFE ($\psi = u - Ts$) and you have the rest!

KKHM began by splitting the HFE into two parts: one at zero density (i.e., rarified gas) plus a second at finite density that vanishes as the density approaches zero. The first part is:

$$\psi_0 = \sum_{i=1}^{8} \frac{C_i}{\tau^{i-1}} + C_7 \ln(T) + C_8 \frac{\ln(T)}{\tau} \quad (A.1)$$

where τ is a dimensionless inverse of the temperature:

$$\tau = \frac{1000}{T} \quad (A.2)$$

where T is the temperature in °K. The whole expression for HFE, ψ, is:

$$\psi = \psi_0 + RT(\ln \rho + \rho Q) \quad (A.3)$$

where ρ is the density in gm/cm³, R is the ideal gas constant (0.46151 J/gm/°K), and Q is the *partition* function, which is:

$$Q = (\tau - \tau_c) \left\{ \sum_{j=1}^{7} (\tau - t_j)^{j-2} \left[\sum_{i=1}^{8} A_{i,j} (\rho - r_j)^{j-1} + e^{-E\rho} \sum_{i=9}^{10} A_{i,j} \rho^{i-9} \right] \right\} \quad (A.4)$$

where $t_j = \tau_c$ for j=1, otherwise $t_j = 2.5$, and rj=0.634 for j=1, otherwise 1. Pressure is given by:

$$p = \rho RT \left(1 + \rho Q + \rho^2 \frac{\partial Q}{\partial \rho} \right) \quad (A.5)$$

Enthalpy is given by:

$$h = RT \left(\rho \tau \frac{\partial Q}{\partial \tau} + 1 + \rho Q + \rho^2 \frac{\partial Q}{\partial \rho} \right) + \frac{d(\psi_0 \tau)}{d\tau} \quad (A.6)$$

Entropy is given by:

$$s = -R\left(\ln \rho + \rho Q - \rho\tau \frac{\partial Q}{\partial \tau}\right) - \frac{d\psi_0}{dT} \quad (A.7)$$

Equations A.1 through A.7 require temperature and density as inputs, yet we most often know temperature and pressure or some other property (saturated liquid, saturated vapor, enthalpy, or entropy). This means that we must solve these lengthy equations implicitly for density over and over again. The precise value of density is rarely of concern; however, an imprecise value of density yields an inaccurate value of pressure, enthalpy, or entropy, which is why we must always obtain an accurate solution.

Solving for the density over the entire range of interest is not as easy as it may seem. The following figure is commonly used to illustrate the problems:

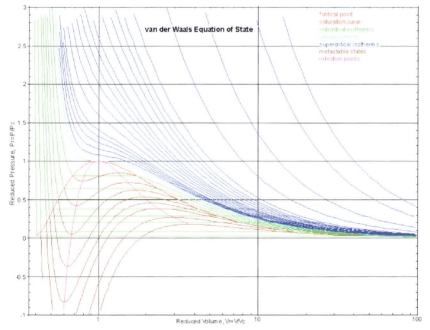

The van der Waals equation of state (i.e., an expression for pressure as a function of temperature and density or specific volume (i.e., 1/density)), $p = RT/(v-b) - a/v^2$, was the first attempt to account for observed fluid behavior. The critical point is where the red, magenta, and cyan lines meet. This is also where both the x- and y-axis equal one. The x-axis is specific volume divided by the value at the critical point and the y-axis is pressure divided by the critical pressure. The cyan curve is the locus of all points where the temperature is equal to the critical temperature. The blue curves are constant temperatures (isotherms) above the critical and the green curves are isotherms below the critical.

The red curve is the saturation line, that is, the states at which the liquid and vapor can coexist in equilibrium. The most common point for water is one atmosphere and 100°C (14.696 psia and 212°F). Notice that the green curves have a discontinuous slope where they intersect the red curve. We will spend a lot of computational time walking up and down this curve! The brown curves are metastable states, that is, they may occur, but they're not in equilibrium and won't last long. This are like water in a pot on the stove at 101°C (213°F) just before a bubble forms and rises to the surface or water in an ice tray in the freezer at -1°C (31°F) just before ice crystals form and clarity disappears.

The green curves on the left side of the critical point are very steep, which can make intersections difficult to calculate. These same curves are very shallow on the far right side, making those intersections difficult to calculate. This is a perfect illustration of nonlinear equation solving and a practical use of the bisection search and Newton-Raphson Method. All the files can be found in the online archive in the folder examples\steam. The main source code is kkhm.c and there is also a batch file to recompile it, although the executable will run on any version of the Windows® operating system.

The first test will be to calculate densities all along the red curve using these two methods. The convergence criterion is the same for both. The results are:

```
timing implicit density calculations
along the saturation line
149,636 function calls
bisection 2.359 seconds
Newton-Raphson 0.415 seconds
```

The bisection search always requires 32 iterations; whereas, the Newton-Raphson typically requires only 2 or 3. The ratio of the run times is 5.68:1, instead of 16:1, because Newton-Raphson requires the pressure and the derivative of pressure with respect to density; whereas, the bisection search does not. The 149,636 cases come from stepping along the red curve from the triple point to the critical point by 0.01°C. Some of the same summations (Equation A.4 and it's derivatives) are used to calculate the pressure and the derivative of the pressure with respect to density and we are careful to reuse these intermediate results for best efficiency. All of the functions are in kkhm.c, including the following snippet:

```
double fQ(double tau,double rho)
  {
  int i,j;
  double dt,q,r,rhoa,s,tauc;
  tauc=1000./Tc;
  dt=1.;
  rhoa=0.634;
  for(q=j=0;j<7;j++)
    {
    r=1.;
```

```
      for(s=i=0;i<8;i++)
        {
        if(j>1&&i>3)
          break;
        s+=A[7*i+j]*r;
        r*=rho-rhoa;
        }
      s+=exp(-E*rho)*(A[7*8+j]+A[7*9+j]*rho);
      q+=dt*s;
      if(j==0)
        dt=tau-tauc;
      else
        dt*=tau-2.5;
      rhoa=1.;
      }
    return(q);
    }
double fP(double tau,double rho)
  {
  double T;
  T=1000./tau;
  return ((fQr(tau,rho)*rho+fQ(tau,rho))*rho+1.)*rho*R*T;
  }
void rPr(double tau,double rho,double*P,double*dPdrho)
  {
  double Q,Qr,Qrr,T;
  T=1000./tau;
  Q=fQ(tau,rho);
  Qr=fQr(tau,rho);
  Qrr=fQrr(tau,rho);
  *P=((Qr*rho+Q)*rho+1.)*rho*R*T;
  *dPdrho=(((Qrr*rho+4.*Qr)*rho+2.*Q)*rho+1.)*R*T;
  }
```

It's also necessary to have some estimate of where to start the search for a root. Because we're always solving the same problem—rather than some general problem—we can use a simple curve fit to supply an estimate of liquid and vapor densities, respectively. Some software packages (including ones I've written) avoid this intensive iterative process by implementing cubic splines. Such are much faster and work reasonably well, except near the critical point, where no such method is adequate.

In fact, some of the tabulated values in the 1969 book are incorrect[48], owing (presumably) to the limited computer resources available at the time. Most the densities between 368°C and 379°C (695°F and 715°F) are off. To six significant figures the critical density is actually 0.316820 gm/cm³ (19.7784

---

[48] When I say "incorrect" I mean that the equations, with the constants provided, don't yield the results listed. They're close, but not close enough.

lbm/ft³), the critical temperature is 647.245°K (374.095° or 705.371°F), and the critical pressure is 22.088 MPa (3203.59 psia). It is essential that the precise (and consistent!) values be used in order to satisfy the following requirement at the critical point:

$$\frac{\partial p}{\partial \rho} = \frac{\partial^2 p}{\partial \rho^2} = 0 \tag{A.8}$$

The next thing we need is a function to return the density anywhere in the preceding figure, not just along the red curve. We will develop one with the bisection search and another using Newton-Raphson and compare the two for speed—requiring the same convergence criteria and that we *always* get a solution. Divergence or failure is *not* an option! There is a test in the code and it will stop should any of the Newton-Raphson function fail to converge.

The first step in finding the density is to determine if we're above or below the cyan curve in the preceding figure. If above, the curves are continuous and we can start searching almost anywhere. If above, the curves are discontinuous and we must limit our search to one side or the other (i.e., liquid or vapor). The curves are very steep on the liquid side (i.e., the left) so that we must not be too far off in our original estimate. The right side (vapor) isn't as much of a problem, but remember Equation A.8 because the first and second partial derivatives are both zero at the critical point. The density locating function using a bisection search is:

```
double frho1(double T,double P) /* rho via bisection */
  {
  int i;
  double Ps,r1,r2,rho,tau;
  if(T<Tc)
    {
    Ps=Psat(T);
    if(P>Ps)
      {
      r1=1.02*frh0f(T);
      r2=P/R/T/2.5;
      }
    else
      {
      r1=P/R/T/2.5;
      r2=1.10*frh0g(T);
      }
    }
  else
    {
    r1=P/R/T/2.5;
    r2=1.1;
    }
  tau=1000./T;
```

```
      for(i=0;i<32;i++)
        {
        rho=sqrt(r1*r2);
        if(fP(tau,rho)<P)
          r1=rho;
        else
          r2=rho;
        }
      return(rho);
      }
```

The functions frh0f() and frh0g() are the approximate values for saturated liquid and vapor densities, respectively. The density function using Newton-Raphson is:

```
    double frho2(double T,double P)  /* rho via Newton-
      Raphson */
    {
    int i;
    double dPdrho,eps,Pp,Ps,rho,tau;
    if(T<Tc)
      {
      Ps=Psat(T);
      if(P>Ps)
        rho=frh0f(T);
      else
        rho=frh0g(T)*P/Ps;
      }
    else
      rho=rhoc*(P/Pc)*(Tc/T);
    if(fabs(T-Tc)>5.)
      eps=1E-6;
    else
      eps=1E-4;
    tau=1000./T;
    for(i=0;i<99;i++)
      {
      rPr(tau,rho,&Pp,&dPdrho);
      if(fabs(Pp/P-1.)<eps)
        return(rho);
      rho+=(P-Pp)/dPdrho;
      }
    fprintf(stderr,"%s(%lG,%lG) failed after %i
      iterations\n",__FUNCTION__,T,P,i);
    exit(1);
    }
```

Results for a series of random values of temperature and pressure are:

```
timing implicit density calculations
away from the saturation line
250,000 function calls
bisection 4.255 seconds
```

```
Newton-Raphson 1.409 seconds
```

Newton-Raphson wins out over bisection search by a factor of 3.02 this time. We will now use these density functions, along with others for enthalpy and entropy, to produce a Mollier Diagram (i.e., a graph of h vs. s), which is very important in the analysis of power systems, especially steam turbines. The figure consists of the saturation line plus a series of isobars (lines of constant pressure) and isotherms (lines of constant temperature). The necessary points are written to an Excel® spreadsheet to produce:

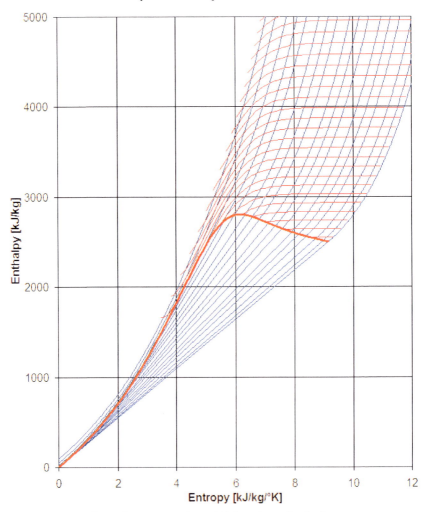

The preceding figure contains 6505 points (and far more property evaluations). The comparative performance is:

```
timing Mollier diagram calculations
```

```
bisection: 0.073 seconds
timing Mollier diagram calculations
Newton-Raphson: 0.028 seconds
```

The Newton-Raphson Method wins out over the bisection search by a factor of 2.6, which is not too impressive, considering the two combined took only a tenth of a second. The file open/close was performed outside the timer loop, but the file writes were inside. You can now see why it's tough to come up with a practical example where it's worth using anything other than a bisection search.

Appendix B: Graphical Representation

There are countless examples of mathematical problems that produce pretty pictures when rendered graphically. Unless you have some software specifically designed to facilitate this, it can be a little problematic. Excel® is not well suited for this task. All you need is a palette and a little code to convert some result into color indices. Write this out to a file and you have a pretty picture. You will find the necessary code in the online archive in folder examples\colormap. A general outline to follow is:

```
int main(int argc,char**argv,char**envp)
   {
   IterationMap(NewtonRaphson,800,800);
   SaveBitmap("Newton-Raphson.bmp");
   return(0);
   }
```

Fill an array with something, as in:

```
void IterationMap(int solver(double,double,double),int
     wide,int high)
   {
   int h,w;
   double guess,target;
   printf("creating iteration map: %ix%i\n",wide,high);
   if((iterations=calloc(wide*high,sizeof(int)))==NULL)
      Abort(__LINE__,"can't allocate memory");
   bi.biWidth=wide;
   bi.biHeight=high;
   for(h=0;h<high;h++)
      {
      guess=-3.+h*6./(high-1);
      for(w=0;w<wide;w++)
         {
         target=0.2+w*0.55/(wide-1);
         iterations[wide*h+w]=solver(target,guess,
   FLT_EPSILON);
         }
      }
   }
```

Linear variation with the number of iterations:

```
   for(h=0;h<bi.biHeight;h++)
      {
      for(w=0;w<bi.biWidth;w++)
         row[w]=(BYTE)(((iterations[bi.biWidth*h+w]-
   mn)*(colors-1))/(mx-mn));
      fwrite(row,1,width,fp);
      }
```

You only need 8 bits/pixel. Don't bother with a JPEG. For a BMP the width in bytes must be a multiple of 4 (i.e., the rows must be DWORD aligned). This will take care of that:

```
bi.biBitCount=8;
width=4*((bi.biWidth*bi.biBitCount+31)/32);
```

Here's what Dekker's method from Chapter 1 looks like:

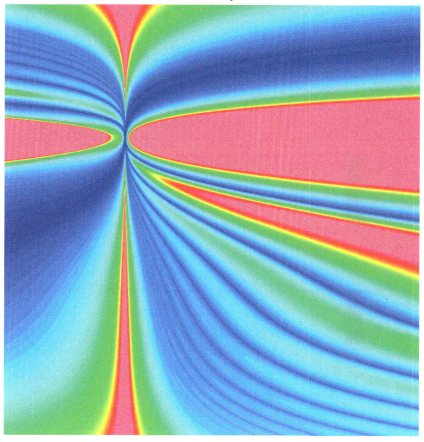

Appendix C: Data Transformation Program

The following program manipulates orthogonal polynomials to arrive at coefficients that will smoothen, interpolate, extrapolate, integrate, and/or differentiate evenly-spaced data in a single step without resorting to matrix inversion. The polynomials P(n,x) have the property that:

$$\sum_{x=0}^{n} P_{nj}(x)P_{nk}(x) \neq 0 \quad j = k$$
$$\sum_{x=0}^{n} P_{nj}(x)P_{nk}(x) = 0 \quad j \neq k \quad (C.1)$$

Milne has shown that these polynomials are given by:[49]

$$P_{nm}(x) = \sum_{i=0}^{m} (-1)^i B(m,i) B(m+i,i) \frac{\Pi(x,i)}{\Pi(n,i)} \quad m = 0,1,2,...,n \quad (C.2)$$

where ***B(i,j)*** is the binomial coefficient and ***Π(n,i)*** is the following product:

$$\Pi(n,i) = n(n-1)(n-2)...(n-i+1) \quad (C.3)$$

The first 3 polynomials are listed below:

$$P_{n0}(x) = 1$$
$$P_{n1}(x) = 1 - \frac{2x}{n} \quad (C.4)$$
$$P_{n2}(x) = 1 - \frac{6x}{n} + \frac{6x(x-1)}{n(n-1)}$$

The code is listed below and may be found in the examples\coefs folder in the on-line archive:

```
/* Determine coefficients that smoothen, interpolate,
      extrapolate, integrate
   and/or differentiate a set of equally-spaced points in
      one step.
   developed by Dudley J. Benton, Knoxville, Tennessee */

#define _CRT_SECURE_NO_DEPRECATE
#include <stdio.h>
#include <stdlib.h>
#include <malloc.h>
#include <float.h>
#define _USE_MATH_DEFINES
#include <math.h>
```

---

[49] Milne, W. E., *Numerical Analysis*, pp. 265-275 and 375-381, Princeton University Press, 1949.

```
void iserror(int test,char*error,int
    line){if(test){fprintf(stderr,"error at line %i:
    %s\n",line,error);exit(1);}}
#define iferror(test) iserror(test,#test,__LINE__)

double binomial(int n,int m)
 { /* binomial coefficient */
 int i,j;
 double b=1.;
 if(n<m||m<0)
 return(0.);
 if(n==m||m<1)
 return(1.);
 if(m>n-m)
 m=n-m;
 for(i=1,j=n;i<=m&&i<j;i++,j--)
 {
 b*=j;
 b/=i;
 }
 return(b);
 }

double polynomial(int n,int m,double x)
 { /* m-th degree polynomial orthogonal over the finite
    set of n points */
 int i;
 double p=1.,q=1.;
 if(m!=0)
 {
 for(i=1;i<=m;i++)
 {
 x-=1;
 n--;
 if(n)
 q*=-x/n;
 p+=binomial(m,i)*binomial(m+i,i)*q;
 }
 }
 return(p);
 }

void derivative(double*C,int p,int d)
 { /* differentiate a polynomial */
 int i,j;
 for(j=1;j<=d;j++)
 {
 for(i=1;i<p;i++)
 C[i-1]=i*C[i];
 C[p-1]=0;
```

```
}
}
void Lagrange(double*A,int n)
{ /* expansion of the Largange polynomial */
int i,j,k,l,new,old;
double*C;
C=calloc(n*2,sizeof(double));
for(i=0;i<n;i++)
{
new=0;
old=n;
C[new]=1;
for(j=1;j<n;j++)
C[j+new]=0;
for(j=l=0;j<n;j++)
{
if(j!=i)
{
l++;
new=old;
old=n-new;
for(k=0;k<l;k++)
C[k+new]=-C[k+old]*(j+1);
C[l+new]=0;
for(k=0;k<l;k++)
C[k+1+new]+=C[k+old];
for(k=0;k<=l;k++)
C[k+new]/=i-j;
}
}
for(j=0;j<n;j++)
A[i*n+j]=C[j+new];
}
free(C);
}

void Smooth(double*S,int p,int s)
{ /* smoothing coefficients for polynomial of degree
    p-1-s */
int i,j,k,l,m;
double*P,*Q,R;
iferror(s<0);
iferror(s>=p);
memset(S,0,p*p*sizeof(double));
if(s<1)  /* if s=0 then [I] */
{
for(i=0;i<p;i++)
S[p*i+i]=1;
return;
```

```c
}
P=calloc(p*p,sizeof(double));
Q=calloc(p*p,sizeof(double));
m=p-s;
for(i=0;i<p;i++)
for(j=0;j<m;j++)
P[p*j+i]=Q[p*i+j]=polynomial(p,j,i+1.);
for(i=0;i<m;i++)
{
for(R=j=0;j<p;j++)
R+=P[p*i+j]*P[p*i+j];
for(j=0;j<p;j++)
P[p*i+j]/=R;
}
for(i=l=0;i<p;i++)
for(j=0;j<p;j++,l++)
for(k=0;k<m;k++)
S[l]+=Q[p*i+k]*P[p*k+j];
free(P);
free(Q);
}

void Coefficients(double*C,int p,int d,int s,double i)
{
int j,k,l;
double*A,*B,*D,*S;
iferror(p<1);
iferror(p>20);
iferror(d<-1);
iferror(s<0);
iferror(s>=p);
iferror(d>0&&s+d>=p);
memset(C,0,p*sizeof(double));
A=calloc(p*p,sizeof(double));
B=calloc(p,sizeof(double));
D=calloc(p,sizeof(double));
S=calloc(p*p,sizeof(double));
Lagrange(A,p);
if(d<0)
{
B[0]=i;
for(k=1;k<p;k++)
B[k]=B[k-1]*i;
for(k=0;k<p;k++)
B[k]=(B[k]-1)/(k+1);
for(k=0;k<p;k++)
for(j=0;j<p;j++)
D[k]+=A[p*k+j]*B[j];
if(fabs(i-1.)>FLT_EPSILON)
for(k=0;k<p;k++)
```

```
      D[k]/=fabs(i-1.);
      }
      else
      {
      B[0]=1;
      for(k=1;k<p;k++)
      B[k]=B[k-1]*i;
      if(d>0)
      for(k=0;k<p;k++)
      derivative(A+p*k,p,d);
      for(k=l=0;k<p;k++)
      for(j=0;j<p;j++,l++)
      D[k]+=A[l]*B[j];
      }
      if(s>0)
      {
      Smooth(S,p,s);
      for(k=l=0;k<p;k++)
      for(j=0;j<p;j++,l++)
      C[k]+=S[l]*D[j];
      }
      else
      memcpy(C,D,p*sizeof(double));
      free(A);
      free(B);
      free(D);
      free(S);
      }

int main(int argc,char**argv,char**envp)
  {
  int d,j,p,s;
  double C[20],i;
  printf("transformation coefficients for equally-spaced
     data\n");
  printf("p=points, i=point, d=differentiate (-1 for
     integrate),s=smoothing\n");
  if(argc!=5)
  {
  printf("usage: coefs p i d s as in\ncoefs 6 7 0 1\n");
  return(1);
  }
  p=atoi(argv[1]);
  i=atof(argv[2]);
  d=atoi(argv[3]);
  s=atoi(argv[4]);
  printf("p=%i,i=%lG,d=%i,s=%i\n",p,i,d,s);
  iferror(p<2);
  iferror(p>20);
  iferror(d<-1);
```

```
iferror(d>=p);
iferror(s<0);
iferror(s>=p);
iferror(d>0&&s+d>=p);
iferror(i<-p);
iferror(i>2*p);
Coefficients(C,p,d,s,i);
printf("%lG",C[0]);
for(j=1;j<p;j++)
printf(",%lG",C[j]);
printf("\n");
return(0);
}
```

Appendix D. Newton-Cotes Coefficient Program

The following program calculates the coefficients for any order of Newton-Cotes integration without inverting a matrix. It is a subset of the program presented in Appendix C. The files may be found in the examples\NewtonCotes folder of the on-line archive.

```
/* Coefficients for Newton-Cotes Integration
 developed by Dudley J. Benton, Knoxville, Tennessee */

#define _CRT_SECURE_NO_DEPRECATE
#include <stdio.h>
#include <stdlib.h>
#include <malloc.h>
#include <float.h>
#define _USE_MATH_DEFINES
#include <math.h>

void iserror(int test,char*error,int
    line){if(test){fprintf(stderr,"error at line %i:
    %s\n",line,error);exit(1);}}
#define iferror(test) iserror(test,#test,__LINE__)

void Lagrange(double*A,int n)
 { /* expansion of the Largange polynomial */
 int i,j,k,l,new,old;
 double*C;
 C=calloc(n*2,sizeof(double));
 for(i=0;i<n;i++)
 {
 new=0;
 old=n;
 C[new]=1;
 for(j=1;j<n;j++)
 C[j+new]=0;
 for(j=l=0;j<n;j++)
 {
 if(j!=i)
 {
 l++;
 new=old;
 old=n-new;
 for(k=0;k<l;k++)
 C[k+new]=-C[k+old]*(j+1);
 C[l+new]=0;
 for(k=0;k<l;k++)
 C[k+1+new]+=C[k+old];
 for(k=0;k<=l;k++)
 C[k+new]/=i-j;
 }
 }
```

```
 for(j=0;j<n;j++)
 A[i*n+j]=C[j+new];
 }
 free(C);
 }

void Coefficients(double*C,int p)
 {
 int j,k;
 double*A,*B;
 iferror(p<1);
 memset(C,0,p*sizeof(double));
 A=calloc(p*p,sizeof(double));
 B=calloc(p,sizeof(double));
 Lagrange(A,p);
 B[0]=p;
 for(k=1;k<p;k++)
 B[k]=B[k-1]*p;
 for(k=0;k<p;k++)
 B[k]=(B[k]-1)/(k+1);
 for(k=0;k<p;k++)
 for(j=0;j<p;j++)
 C[k]+=A[p*k+j]*B[j];
 for(k=0;k<p;k++)
 C[k]/=p-1;
 free(A);
 free(B);
 }

int main(int argc,char**argv,char**envp)
 {
 int j,p;
 double*C;
 printf("Newton-Cotes integration coefficients\n");
 if(argc!=2)
 {
 printf("usage: NewtonCotes 6\n");
 return(1);
 }
 p=atoi(argv[1]);
 iferror(p<2);
 C=calloc(p,sizeof(double));
 Coefficients(C,p);
 printf("%lG",C[0]);
 for(j=1;j<p;j++)
 printf(",%lG",C[j]);
 printf("\n");
 return(0);
 }
```

The coefficients for orders up to 16 are listed below. The coefficients up to 60th may be found in the examples\NewtonCotes folder in the on-line archive.

```
double NC2[]={1./2.,1./2.};
double NC3[]={1./6.,2./3.,1./6.};
double NC4[]={1./8.,3./8.,3./8.,1./8.};
double NC5[]={7./90.,32./90.,12./90.,32./90.,7./90.};
double NC6[]={19./288.,75./288.,50./288.,50./288.,
   75./288.,19./288.};
double NC7[]={41./840.,216./840.,27./840.,272./840.,
   27./840.,216./840.,41./840.};
double NC8[]={751./17280.,3577./17280.,1323./17280.,
   2989./17280.,2989./17280.,1323./17280.,3577./17280.,
   751./17280.};
double NC9[]={989./28350.,5888./28350.,-928./28350.,
   10496./28350.,-4540./28350.,10496./28350.,
   -928./28350.,5888./28350.,989./28350.};
double NC10[]={2857./89600.,15741./89600.,1080./89600.,
   19344./89600.,5778./89600.,5778./89600.,
   19344./89600.,1080./89600.,15741./89600.,
   2857./89600.};
double NC11[]={16067./598752.,106300./598752.,
   -48525./598752.,272400./598752.,
   -260550./598752.,427368./598752.,
   -260550./598752.,272400./598752.,
   -48525./598752.,106300./598752.,16067./598752.};
double NC12[]={0.0249332309119641,0.1548553585207226,
   -0.0371692317937990,0.2896582547949741,
   -0.1101780891754860,0.1779004767416227,
   0.1779004767416230,-0.1101780891754863,
   0.2896582547949741,-0.0371692317937991,
   0.1548553585207226,0.0249332309119641};
double NC16[]={0.0170872997716264,0.1285073786774053,
   -0.1127229050595076,0.5070427082102271,
   -0.7562931148441474,1.1913603495068070,
   -0.9680052114962560,0.4930234952338425,
   0.4930234952338438,-0.9680052114962570,
   1.1913603495068090,-0.7562931148441480,
   0.5070427082102273,-0.1127229050595076,
   0.1285073786774054,0.0170872997716265};
```

Appendix E. Gauss Quadrature Weights & Abscissas

As discussed in Chapter 15, the abscissas for Gauss Quadrature (GQ) are the roots of the n+1 degree Legendre Polynomial and the weights are values of the nth degree polynomial at each of the roots. The abscissas are found by locating the roots and the weights are found by plugging these values into the appropriate polynomial. The following program accomplishes this:

```
/* Find coefficients (weights and abscissas) for Gauss
   Quadrature
   This version uses extended-precision floating-point
   operations.
   developed by Dudley J. Benton, Knoxville, Tennessee
   */
#define _CRT_SECURE_NO_DEPRECATE
#include <stdio.h>
#include <stdlib.h>
#include <string.h>
#include <malloc.h>
#include <float.h>
#define _USE_MATH_DEFINES
#include <math.h>
#include "xreal.hpp"
#include "real.cpp"

real tiny;

void LegendrePolynomial(int n,real X,real*P3,real*Q3)
  {
  int i;
  real P1,P2,Q1,Q2;
  *P3=1.;
  *Q3=0.;
  if(n<1)
    return;
  P2=*P3;
  Q2=*Q3;
  *P3=X;
  *Q3=1.;
  if(n<2)
    return;
  for(i=2;i<=n;i++)
    {
    P1=P2;
    Q1=Q2;
    P2=*P3;
    Q2=*Q3;
    *P3=2.*X*P2-P1-(X*P2-P1)/i;
    *Q3=Q1+(2*i-1)*P2;
    }
  }
```

```
real RefineRoot(int n,real X1,real X2)
  {
  int i;
  real dP,P,Xo;
  static real X;
  X=(X1+X2)/2.;
  for(i=0;i<128;i++)
    {
    LegendrePolynomial(n,X,&P,&dP);
    if(fabs(dP)<0.5)
      break;
    Xo=X;
    X=fmax(X1,fmin(X2,X-P/dP));
    if(fabs(P)<=tiny||fabs(X-Xo)<=tiny)
      return(X);
    }
  X=2.;
  return(X);
  }

int InsertRoot(real*Roots,int nr,real Root)
  {
  int i,j;
  for(i=0;i<nr;i++)
    if(fabs(Roots[i]-Root)<tiny)
      return(0);
  for(i=1;i<nr;i++)
    if(Roots[i-1]<Root&&Root<Roots[i])
      break;
  for(j=nr;j>i;j--)
    Roots[j]=Roots[j-1];
  Roots[i]=Root;
  return(i);
  }

int FindRoots(int n,real*Roots,real*Weights)
  {
  int i,j,k,nr;
  real dP,dX,P,Root,X,X1,X2;
  printf("computing coefficients\n");
  if(n<2)
    return(0);
  nr=1;
  if(n%2)
    Roots[0]=0;
  else
    {
    X1=0.;
    X2=1.3/pow(n,0.94);
```

```
      Roots[0]=RefineRoot(n,X1,X2);
      if(n<3)
        goto weights;
      }
   X1=1.-2.1/pow(n,1.9);
   X2=1.;
   Roots[nr++]=RefineRoot(n,X1,X2);
   if(n<5)
     goto weights;
   printf("\r%i roots found out of %i",nr,(n+1)/2);
   while(nr<(n+1)/2)
      {
      for(i=1;i<nr;i++)
         {
         k=(n+1)/2-nr+1;
         dX=(Roots[i]-Roots[i-1])/k;
         X2=Roots[i-1];
         for(j=0;j<k;j++)
            {
            X1=X2;
            X2=X2+dX;
            Root=RefineRoot(n,X1,X2);
            if(Root>1.5)
               continue;
            if(!InsertRoot(Roots,nr,Root))
               continue;
            nr++;
            printf("\r%i roots found out of
    %i",nr,(n+1)/2);
            break;
            }
         if(nr>=(n+1)/2)
            goto weights;
         }
      }
   weights:
   if(n>4)
     printf("\n");
   for(i=0;i<(n+1)/2;i++)
      {
      X=Roots[i];
      LegendrePolynomial(n,X,&P,&dP);
      Weights[i]=2/(1-X*X)/dP/dP;
      }
   return((n+1)/2);
   }

int main(int argc,char**argv,char**envp)
   {
   int i,n;
```

```
real*A,*W;
printf("Gauss Quadrature Weights and Abscissas\n");
if(argc!=2)
   {
   printf("usage: GQUAD 123\n");
   return(1);
   }
n=atoi(argv[1]);
if(n<2)
   {
   printf("n must be at least 2\n");
   return(1);
   }
A=(real*)calloc(((n+1)/2),sizeof(real));
W=(real*)calloc(((n+1)/2),sizeof(real));
tiny=DBL_EPSILON*DBL_EPSILON;
FindRoots(n,A,W);
printf("abscissas\n");
for(i=0;i<(n+1)/2;i++)
   printf("%s\n",xtoa(&A[i].x));
printf("weights\n");
for(i=0;i<(n+1)/2;i++)
   printf("%s\n",xtoa(&W[i].x));
return(0);
}
```

The roots up to 4096 may be found in the examples\GaussQuadrature folder in the on-line archive. The weights and abscissas for orders 3 through 6 were listed in Chapter 15. The weights and abscissas for orders 7 through 12 are listed below:

```
double A7[]={0.000000000000,0.405845151377,
   0.741531185599,0.949107912343};
double W7[]={0.417959183673,0.381830050505,
   0.279705391489,0.129484966169};
double A8[]={0.183434642496,0.525532409916,
   0.796666477414,0.960289856498};
double W8[]={0.362683783378,0.313706645878,
   0.222381034453,0.101228536290};
double A9[]={0.000000000000,0.324253423404,
   0.613371432701,0.836031107327,0.968160239508};
double W9[]={0.330239355001,0.312347077040,
   0.260610696403,0.180648160695,0.081274388362};
double A10[]={0.148874338982,0.433395394129,
   0.679409568299,0.865063366689,0.973906528517};
double W10[]={0.295524224715,0.269266719310,
   0.219086362516,0.149451349151,0.066671344309};
double A12[]={0.125233408511,0.367831498998,
   0.587317954287,0.769902674194,0.904117256370,
   0.981560634247};
double W12[]={0.249147045813,0.233492536538,
```

```
  0.203167426723,0.160078328543,0.106939325995,
  0.047175336387};
```

The odd orders are not as efficient as the even ones, which is why you rarely see the high order odd coefficients.

Static vs. Global Data in C

Programming quadrature often requires large data statements, something you can't do in VBA® and is really convoluted in VB.Net®. Data statements are natural structures in all other programming languages, which is yet another reason to graduate from BASIC to a professional language.[50] In C, all variables with global scope (i.e., outside a function) are inherently static. In spite of this fact, you often see code and/or data with the qualifier "static" at the global level. This indicates that whoever wrote the code doesn't understand C or how compilers work, as this qualifier is ignored in this context.

At the local level (i.e., within the context of a function), all variables are *discardable* unless qualified as *static*. All discardable objects are given temporary locations on the stack.[51] This means that in the code below:

```
function MyQuad(double f(double x),double a,double b)
  {
  int i;
double q;
double A[4]={0.,0.4,0.6,1.0};
double W[4]={0.25,0.25,0.25,0.25};
for(q=i=0;i<4;i++)
  q+=W[i]*f(a+A[i]*(b-a));
return(q);
}
```

the statements A[4]={... and W[4]={... will be *executed* each and every time you enter the function! If you look at the assembler (i.e., machine language instructions) that the compiler generates, you will see the constants stored somewhere in global memory are copied into the temporary locations on the stack every time. This may take as long as the function itself. If you mean for these constants to always be available and have these same values within the context of a function, then qualify them with static, as in:

```
static double A[4]={0.,0.4,0.6,1.0};
static double W[4]={0.25,0.25,0.25,0.25};
```

---

[50] Never forget that BASIC stands for *Beginner's* All-Purpose Symbolic Instruction Code. It's the programmer's equivalent of Pull-Ups®, the popular transition from diapers to potty training.

[51] Intel® processors are *stack* machines. Most modern processors are. The *stack* is a designated block of memory that facilitates calling and returning from procedures, passing parameters (i.e., arguments), and temporary storage. It's called a *stack* because it's most often accessed like cafeteria trays: last on is first off, first on is last off.

Remember… you only need to do this at the local level.

Extended Precision

You may notice that some of the weights and abscissas contained in the examples exceed that which can be produced with traditional code. That's because I ran those out specifically for this purpose using extended precision. The Microsoft® C compilers of the past decade no longer support 80-bit floating-point numbers, but Intel processors always have and still do. I still have a copy of the old Microsoft® C compiler that does. Of course, it produces 16-bit executables that won't run on a 64-bit O/S except in a virtual box. Even 80-bit only gets you so far.

There is another way… extended precision. Several sources of extended precision math code may be found on the Web. The ones I have been able to find are only compatible with the GNU compiler, gcc, which less common than three-headed talking chickens and far less useful. Pass one of these codes to the Microsoft® C compiler and your machine will self-destruct in a cry of agony.

Fortunately for you, I have developed my own extended precision math code, which you are welcome to have and use. Here's how it works…

```
/* define extended precision real functions
   number  digits
     of      of
   bytes  precision
     8      15  <- minimum
    12      25
    16      35    note: set the number of bytes for
    20      44          the xreal in the typedef below
    24      54
    28      63    note: the number of bytes is not
    32      73          limited to multiples of 4
    36      82
    40      92
    44     102
    48     111
    52     121
    56     131
    60     140
    64     150 */
typedef struct xreal{BYTE b[24];}xreal;
```

You will find all of these files in the xreal folder. This definition is in file xreal.hpp. IEEE 754 double precision is 64-bit or 8 bytes. As shown in the preceding table, this yields approximately 15 significant digits. If you dimension b[8] in the xreal structure, the code will just use the FPU and be a whole lot faster. If you dimension b[12], b[16], etc., extended precision will take over.

The first 8 bytes of xreal are exactly the same as IEEE 754 or Intel® 64-bit double precision floating-point numbers. The bytes beyond 8 are additional mantissa. The exponent part is identical for xreal of any size. This is different from the Intel® 80-bit numbers that have exponents of 2^{-1024} to 2^{+1023} or about 10^{-308} to 10^{+308}. The header file xreal.hpp defines a variety of functions, as listed below. The code to implement these calculations is in xreal.cpp.

char*	xtoa(xreal*)	encode an xreal
char*	xtohex(xreal*x)	xreal to hex
int	intx(xreal*)	convert xreal to int
int	nintx(xreal*)	convert xreal to nearest int
int	xalmost(xreal*x,xreal*y)	xreals almost the same?
int	xclose(xreal*x,xreal*y)	xreals close?
int	xcmp(xreal*,xreal*)	compare 2 xreals
int	xcmpz(xreal*x)	compare xreal to zero
int	xisinf(xreal*x)	is?inf
int	xisnan(xreal*x)	is?nan
int	xiszero(xreal*)	test for not zero
int	xnearly(xreal*x,xreal*y)	xreals nearly the same?
int	xsame(xreal*,xreal*)	compare 2 xreals
double	xtod(xreal*)	convert xreal to double
xreal	atox(char*)	decoded an xreal
xreal	dtox(double d)	convert double to xreal
xreal	itox(int i)	convert int to xreal
xreal	xabs(xreal*)	absolute value
xreal	xacos(xreal*)	arccos(x)
xreal	xadd(xreal*,xreal*)	add 2 xreals
xreal	xasin(xreal*)	arcsin(x)
xreal	xatan(xreal*)	arctan(x)
xreal	xatan2(xreal*,xreal*)	arctan(x/y)
xreal	xchs(xreal*)	change signs
xreal	xcos(xreal*)	cos(x)
xreal	xdiv(xreal*,xreal*)	divide 2 xreals
xreal	xdot(xreal*p,xreal*q,int n)	$d=sum(p[i]*q[i]), i=1..n$
xreal	xe()	return exp(1)
xreal	xexp(xreal*)	return exp(x)
xreal	xhalf(xreal*x)	divide an xreal by 2
xreal	xinf()	inf
xreal	xlarge()	largest possible xreal
xreal	xlog(xreal*)	return ln(x)
xreal	xlog10(xreal*)	return log(x)
xreal	xmax(xreal*x,xreal*y)	max(x,y)
xreal	xmin(xreal*x,xreal*y)	min(x,y)
xreal	xmul(xreal*,xreal*)	multiply 2 xreals
xreal	xnan()	nan

xreal	xone()	return 1
xreal	xpi()	return π
xreal	xpow(xreal*,xreal*)	power x^y
xreal	xsin(xreal*)	sin(x)
xreal	xsmall()	smallest possible xreal >0
xreal	xsqrt(xreal*)	square root of xreal
xreal	xsub(xreal*,xreal*)	subtract 2 xreals
xreal	xtan(xreal*)	tan(x)
xreal	xten()	return 10
xreal	xtoi(xreal*,int)	raise xreal to int
xreal	xtwice(xreal*x)	multiply an xreal by 2
xreal	xtwo()	return 2
xreal	xzero()	zero
void	xset(xreal*p,int n,xreal*q)	set
void	xcopy(xreal*p,xreal*q,int n)	copy

The following code adds two extended floating-point numbers:

```
xreal xadd(xreal*x,xreal*y) /* add two xreals */
  {
  BYTE mask;
  int i,j,k,xe,xs,ye,ys,ze,zs;
  xreal xm,ym,zm;
  static xreal s;
  union{WORD w;BYTE b[2];}exponent,u;

  if(sizeof(xreal)==8)
    {
    xd.d=*(double*)x;
    xd.d+=*(double*)y;
    return(xd.x);
    }
  if(xiszero(y)) /* if y=0 then return x */
    return(*x);
  if(xiszero(x)) /* if x=0 then return y */
    return(*y);

  s=*x;
  exponent.b[1]=(s.b[sizeof(xreal)-1]>>4)&0x07; /*
    unpack xreal */
  exponent.b[0]=(s.b[sizeof(xreal)-
    1]<<4)|(s.b[sizeof(xreal)-2]>>4);
  xe=exponent.w-1023-(sizeof(xreal)*8-11);
  xm=xzero();
  for(i=0;i<sizeof(xreal)-1;i++)
    xm.b[i]=s.b[i];
  xm.b[i-1]&=0x0F;
  xm.b[i-1]|=0x10;
```

```
xs=xisneg(&s);

exponent.b[1]=(y->b[sizeof(xreal)-1]>>4)&0x07;  /*
  unpack xreal */
exponent.b[0]=(y->b[sizeof(xreal)-1]<<4)|(y-
  >b[sizeof(xreal)-2]>>4);
ye=exponent.w-1023-(sizeof(xreal)*8-11);
ym=xzero();
for(i=0;i<sizeof(xreal)-1;i++)
  ym.b[i]=y->b[i];
ym.b[i-1]&=0x0F;
ym.b[i-1]|=0x10;
ys=xisneg(y);

if(ye>xe) /* make x the one with the larger exponent
  */
  {
  zs=xs;
  ze=xe;
  zm=xm;
  xs=ys;
  xe=ye;
  xm=ym;
  ys=zs;
  ye=ze;
  ym=zm;
  s=*y;
  }

if(xe-ye>sizeof(xreal)*8-11)  /* If the exponents
  differ by more than the */
  return(s); /* number of significant digits don't
  bother adding them. */

if(xs) /* if x<0 fix signed integer */
  {
  for(i=0;i<sizeof(xreal);i++)
    xm.b[i]^=0xFF; /* The negative of a number is
  found by */
  u.b[1]=0; /* XORing it with -1 and adding 1. */
  u.b[0]=xm.b[0];
  u.w+=1;
  xm.b[0]=u.b[0];
  for(i=1;i<sizeof(xreal);i++) /* add with carry */
    {
    u.b[0]=u.b[1];
    u.b[1]=0;
    if(u.w==0)
      break;
    u.w+=xm.b[i];
```

```
      xm.b[i]=u.b[0];
      }
   }

if(ys)  /* if y<0 fix signed integer */
   {
   for(i=0;i<sizeof(xreal);i++)
     ym.b[i]^=0xFF; /* The negative of a number is
  found by */
   u.b[1]=0; /* XORing it with -1 and adding 1. */
   u.b[0]=ym.b[0];
   u.w+=1;
   ym.b[0]=u.b[0];
   for(i=1;i<sizeof(xreal);i++) /* add with carry */
      {
      u.b[0]=u.b[1];
      u.b[1]=0;
      if(u.w==0)
        break;
      u.w+=ym.b[i];
      ym.b[i]=u.b[0];
      }
   }

while(ye<xe) /* shift y to the right until the two
  exponents are equal */
   {
   if(xe-ye>=8) /* shift one BYTE at a time */
      {
      for(i=0;i<sizeof(xreal)-1;i++)
        ym.b[i]=ym.b[i+1];
      if(ys)
        ym.b[i]=0xFF;
      ye+=8;
      }
   else /* shift one bit at a time */
      {
      for(i=0;i<sizeof(xreal)-1;i++)
         {
         ym.b[i]>>=1;
         if(ym.b[i+1]&1)
           ym.b[i]|=0x80;
         }
      ym.b[i]>>=1;
      if(ys)
        ym.b[i]|=0x80;
      ye++;
      }
   }
```

```
  u.b[1]=0; /* add the two integers */
  for(i=0;i<sizeof(xreal);i++)
     {
     u.b[0]=u.b[1];
     u.b[1]=0;
     u.w+=xm.b[i];
     u.w+=ym.b[i];
     xm.b[i]=u.b[0];
     }

  if(xisneg(&xm)) /* convert to positive integer */
     {
     xs=1;
     for(i=0;i<sizeof(xreal);i++)
        xm.b[i]^=0xFF;
     u.b[1]=0;
     u.b[0]=xm.b[0];
     u.w+=1;
     xm.b[0]=u.b[0];
     for(i=1;i<sizeof(xreal);i++)
        {
        u.b[0]=u.b[1];
        u.b[1]=0;
        if(u.w==0)
          break;
        u.w+=xm.b[i];
        xm.b[i]=u.b[0];
        }
     }
  else
     xs=0;

  k=-1; /* determine position of first non-zero bit of
     result */
  j=8*sizeof(xreal);
  for(i=sizeof(xreal)-1;i>=0&&k<0;i--)
     {
     for(mask=0x80;mask;mask>>=1,j--)
        {
        if(xm.b[i]&mask)
           {
           k=j;
           break;
           }
        }
     }

  if(k<0) /* check for zero result (underflow) */
     {
     s=xzero();
```

```
  return(s);
  }
while(k>sizeof(xreal)*8-11) /* shift result right
  until normalized */
  {
  if(k-(sizeof(xreal)*8-11)>=8) /* shift one BYTE at
  a time */
    {
    for(i=0;i<sizeof(xreal)-1;i++)
      xm.b[i]=xm.b[i+1];
    xm.b[i]=0x00;
    xe+=8;
    k-=8;
    }
   else /* shift one bit at a time */
    {
    for(i=0;i<sizeof(xreal)-1;i++)
      {
      xm.b[i]>>=1;
      if(xm.b[i+1]&0x01)
        xm.b[i]|=0x80;
      }
    xm.b[i]>>=1;
    xe++;
    k--;
    }
  }

while(k<sizeof(xreal)*8-11) /* shift result left
  until normalized */
  {
  if((sizeof(xreal)*8-11)-k>=8) /* shift one BYTE at
  a time */
    {
    for(i=sizeof(xreal)-1;i>0;i--)
      xm.b[i]=xm.b[i-1];
    xm.b[i]=0;
    xe-=8;
    k+=8;
    }
   else /* shift one bit at a time */
    {
    for(i=sizeof(xreal)-1;i>0;i--)
      {
      xm.b[i]<<=1;
      if(xm.b[i-1]&0x80)
        xm.b[i]|=1;
      }
    xm.b[i]<<=1;
```

```
      xe--;
      k++;
      }
    }

  xm.b[sizeof(xreal)-2]&=0x0F;  /* copy result to x */
  exponent.w=(WORD)(xe+1023+(sizeof(xreal)*8-11));
  xm.b[sizeof(xreal)-2]|=(exponent.b[0]<<4);
  xm.b[sizeof(xreal)-
    1]=(exponent.b[1]<<4)|(exponent.b[0]>>4);
  s=xm;
  if(xs)  /* set sign bit */
    s.b[sizeof(xreal)-1]|=0x80;

  return(s);
  }
```

You will find no copyright notice in these files other than my name, because the operations are not based on the work of anyone else. I first developed these operations in assembler to run on the HP-1000 minicomputer at the TVA Engineering Laboratory in Norris around 1982. I published them in 1992 after translating them into Intel® assembler.

While you can use these directly, you don't have to. I have also provided a C++ wrapper in the form of type real. The xreal is inside the class real. The class real facilitates access to the xreal operations that can be used just like common double precision ones. There are also some predefined constants, including: pi, e, gamma, golden, and ln(10).

The real class is as follows, which shows the supported operators:

```
class real
  {
  public:
  inline real d2r(double d){x=dtox(d);return(*this);}
  real  operator = (double d){return d2r(d);};
  friend real fabs(real);
  friend real fchs(real);
  friend real sqrt(real);
  friend real exp(real);
  friend real log(real);
  friend real pow(real,real);
  friend int operator  <  (real,real);
  friend int operator  <  (real,double);
  friend int operator  >  (real,real);
  friend int operator  >  (real,double);
  friend int operator != (real,real);
  friend int operator <= (real,real);
  friend int operator == (real,real);
  friend int operator >= (real,real);
  friend real operator *  (real,real);
  friend real operator *  (real,double);
```

```
    friend real operator * (double,real);
    friend real operator * (int,real);
    friend real operator * (real,int);
    friend real operator + (real,real);
    friend real operator + (real,double);
    friend real operator + (double,real);
    friend real operator - (real,real);
    friend real operator - (real,double);
    friend real operator - (double,real);
    friend real operator / (real,real);
    friend real operator / (real,double);
    friend real operator / (double,real);
    friend real operator / (real,int);
    xreal x;
    };
```

The results for 20-pt Gauss Quadrature are:

```
gquad 20
Gauss Quadrature Weights and Abscissas
computing coefficients
10 roots found out of 10
abscissas
+7.6526521133497333754640409398838211004796266813497499827E-
    002
+2.2778585114164507808049619536857462474308893768292747007E-
    001
+3.7370608871541956067254817702492723739574632170568270822E-
    001
+5.1086700195082709800436405095525099842549132920242682777E-
    001
+6.3605368072651502545283669622628593674338911679936845644E-
    001
+7.4633190646015079261430507035564159031073067956917643545E-
    001
+8.3911697182221882339452906170152068532962936506563736371E-
    001
+9.1223442825132590586775244120329811304918479742369176473E-
    001
+9.6397192727791379126766613119727722191206032780618884505E-
    001
+9.9312859918509492478612238847132027822264713090165588633E-
    001
weights
+1.5275338713072585069808433195509759349194864511237859526E-
    001
+1.4917298647260374678782873700196943669267990408136831628E-
    001
+1.4209610931838205132929832506716493303451541339202030150E-
    001
+1.3168863844917662689849449974816313491611051114698352684E-
    001
+1.1819453196151841731237737771138228700504121954896877299E-
    001
+1.0193011981724043503675013548034987616669165602339255654E-
    001
```

```
+8.32767415767047487247581432220462061001778285831632 89241E-
    002
+6.26720483341090635695065351870416063516010765784363 64211E-
    002
+4.06014298003869413310399522749321098790906399899515 35539E-
    002
+1.76140071391521183118619623518528163621431055433367 33324E-
    002
```

Orders up to 4096 may be found in the \examples\GaussQuadrature folder.

Appendix F. Cooling Tower Demand

As mentioned in Chapters 15 and 21, cooling tower demand is calculated using numerical integration. The following program not only illustrates 3 different calculation methods (4-pt Chebyshev, 10-pt Lobatto, 4th order Runge-Kutta), but also includes a variety of moist air property function required for these calculations:

```c
/* The purpose of this program is to compute KaV/L
   using 3 methods
   This program was developed by Dudley J. Benton */

#define _CRT_SECURE_NO_DEPRECATE
#include <stdio.h>
#include <stdlib.h>
#include <float.h>
#define _USE_MATH_DEFINES
#include <math.h>

/* thermodynamic constants */

double Patm =14.696  ; /* change this and the
   properties will change accordingly */
double Cpa  =0.2406  ; /* constant pressure specific
   heat of dry air */
double Cpg  =0.427933; /* temperature variation of
   water vapor enthalpy */
double Cpw  =1.0;      ; /* constant pressure specific
   heat of water */
double Two  =32.018  ; /* triple point of water */
double Hwo  =1061.39 ; /* enthalpy of water vapor at
   0øF */
double Rmw  =0.62198 ; /* ratio of water to air
   molecular weights */
double Lewis=1.2     ; /* Lewis Number */

/* moist air properties based on Hyland & Wexler */

double fs(double T,double P)
  {
  return(((-4.55447E-10*T+9.400757E-09*P+1.282159E-
    07)*T-1.762686E-06*P+6.35199E-06)*T+3.18886E-
    04*P+1.000104);
  }

double fPwsat(double T)
  {
  double P1=-3.99087,P2=1.81071E-1,P3=-3.63953E-
    3,P4=2.7966E-5,
      P5=-6.73798E-8,P6=9.27329E-11,Q1=-3.22286E-
    2,Q2=4.61283E-4;
```

```
   return(exp((((((P6*T+P5)*T+P4)*T+P3)*T+P2)*T+P1)/((Q
   2*T+Q1)*T+1.))*fs(T,Patm));
   }

double fTwsat(double P)
   {
   double
   P1=101.729,P2=95.4163,P3=33.8827,P4=5.24155,P5=0.347
   441,

   P6=0.0167944,P7=0.000362087,Q1=0.609043,Q2=0.110268,
   T;
   P=log(P);

   T=((((((P7*P+P6)*P+P5)*P+P4)*P+P3)*P+P2)*P+P1)/((Q2*
   P+Q1)*P+1.);
   P-=log(fs(T,Patm));

   T=((((((P7*P+P6)*P+P5)*P+P4)*P+P3)*P+P2)*P+P1)/((Q2*
   P+Q1)*P+1.);
   return(T);
   }

double fWsat(double T)
   {
   double P;
   P=fPwsat(T);
   return(Rmw*P/(Patm-P));
   }

double fPvw(double W) /* partial pressure of water
   vapor from humidity ratio */
   {
   return(W*Patm/(Rmw+W));
   }

double fHdbw(double Tdb,double W)
   {
   return(Cpa*Tdb+W*(Hwo+Cpg*Tdb));
   }

double fHtwb(double Twb)
   {
   return(fHdbw(Twb,fWsat(Twb)));
   }

double Ftwbh(double H)
   {
   int i;
```

```
double Twb,Twb1,Twb2;
Twb1=-30.;
Twb2=fTwsat(Patm)-1.;
for(i=0;i<25;i++)
   {
   Twb=(Twb1+Twb2)/2.;
   if(fHtwb(Twb)<H)
      Twb1=Twb;
   else
      Twb2=Twb;
   }
return(Twb);
}

double fWdwb(double Tdb,double Twb)
   {
   double X,Y;
   if(Tdb<=Twb)
      return(fWsat(Tdb));
   X=(Hwo+Cpw*(Two-Twb)+Cpg*Twb)*fWsat(Twb)-Cpa*(Tdb-
   Twb);
   Y=Hwo+Cpw*(Two-Twb)+Cpg*Tdb;
   return(X/Y);
   }

double fPvdwb(double Tdb,double Twb)
   {
   return(fPvw(fWdwb(Tdb,Twb)));
   }

double fHdwb(double Tdb,double Twb)
   {
   return(fHdbw(Tdb,fWdwb(Tdb,Twb)));
   }

double fTdbhw(double H,double W)
   {
   return((H-W*Hwo)/(Cpa+W*Cpg));
   }

double fRHdwb(double Tdb,double Twb)
   {
   if(Tdb<=Twb)
      return(100.);
   return(100.*fPvdwb(Tdb,Twb)/fPwsat(Tdb));
   }

double fTdbwrh(double Twb,double Rh)
   {
   int i;
```

```
  double Tdb,Tdb1,Tdb2;
  if(Rh>=100.)
    return(Twb);
  Tdb1=Twb;
  Tdb2=fHdwb(Twb,Twb)/Cpa;
  for(i=0;i<25;i++)
    {
    Tdb=(Tdb1+Tdb2)/2.;
    if(fRHdwb(Tdb,Twb)>Rh)
      Tdb1=Tdb;
    else
      Tdb2=Tdb;
    }
  return(Tdb);
  }

double KMerkel(double Twb,double Ran,double App,double
    LG)
  { /* KaV/L from Merkel's equation using 4-point
    Chebyshev */
  int i;
  double asym,dHa,Ha,Hai,Hao,Hw,KaV,Tco,Tw;
  static double a[4]={.10,.40,.60,.90};
  static double w[4]={.25,.25,.25,.25};
  Tco=Twb+App;
  Hai=fHtwb(Twb);
  Hao=fHtwb(Tco+.9*Ran);
  asym=(Hao-Hai)/.9/Ran/Cpw;
  if(LG<0.)
    return(asym);
  if(LG>=asym)
    return(99.99);
  dHa=Ran*Cpw*LG;
  KaV=0.;
  for(i=0;i<4;i++)
    {
    Ha=Hai+a[i]*dHa;
    Tw=Tco+a[i]*Ran;
    Hw=fHtwb(Tw);
    if(Hw<=Ha)
      return(99.99);
    KaV=KaV+w[i]/(Hw-Ha);
    }
  KaV=Cpw*Ran*KaV;
  return(KaV);
  }

double KMarcel(double Twb,double Ran,double App,double
    LG)
```

```
{ /* KaV/L from Marcel Lefevre's equation using 10-
  point Lobatto */
int i;
double
  asym,Ha,Hai,Hao,KaV,Ta,Tai,Tao,Tao1,Tao2,Tco,Tho,Tw,
  W,Wai,Wao,Ws;
static double
  a[10]={.000000,.040233,.130613,.261038,.417361,.5826
  39,.738962,.869387,.959767,1.00000};
static double
  w[10]={.011111,.066653,.112445,.146021,.163770,.1637
  70,.146021,.112445,.066653,.011111};
Tco=Twb+App;
Tho=Tco+Ran;
Tai=Twb;
Hai=fHtwb(Tai);
Wai=fWsat(Tai);
Tao=Tho;
Hao=fHtwb(Tao);
Wao=fWsat(Tao);
asym=((Hao-Hai)/Cpw-(Wao-Wai)*(Tco-Two))/Ran;
if(LG<=0.)
  return(asym);
if(LG>=asym)
  return(99.99);
Tao1=Twb;
Tao2=Tho;
for(i=0;i<25;i++)
   {
   Tao=(Tao1+Tao2)/2.;
   Hao=fHtwb(Tao);
   Wao=fWsat(Tao);
   if(Hao<Hai+Cpw*(LG*Ran+(Wao-Wai)*(Tco-Two)))
     Tao1=Tao;
   else
     Tao2=Tao;
   }
KaV=0.;
for(i=0;i<10;i++)
   {
   Ha=Hao+a[i]*(Hai-Hao);
   Ta=Ftwbh(Ha);
   W=fWsat(Ta);
   Tw=(Cpw*(LG*Tho-(Wao-W)*Two)-Hao+Ha)/Cpw/(LG-
   Wao+W);
   Ws=fWsat(Tw);
   if(Tw<=Ta)
     return(99.99);
   KaV=KaV+w[i]/(Lewis*Cpa*(Tw-Ta)+(Hwo+Cpg*Ta)*(Ws-
   W)/(1.+W));
```

```
    }
  KaV=Cpw*Ran*KaV;
  return(KaV);
  }

double KExact(double Tdb,double Twb,double Ran,double
    App,double LG)
  { /* KaV/L from exact equations using 4th-order
    Runge-Kutta */
  int i,j,k;
  double
    asym,H,Ha,dH[4],Hai,Hao,KaV,K1,K2,LG1,LG2,Ta,Tco,Tho
    ,Tw,W,Wa,dW[4],Wai,Wao,Ws,dY;
  double A[3]={.5,.5,1.}; /* 4th-order Runge-Kutta */
  double B[4]={1./6,1./3,1./3,1./6}; /* coefficients */
  int n=10; /* number of steps to integrate */
  Tco=Twb+App; /* compute asymptote */
  Tho=Tco+Ran;
  Hai=fHdwb(Tdb,Twb);
  Wai=fWdwb(Tdb,Twb);
  Hao=fHtwb(Tho);
  Wao=fWsat(Tho);
  asym=((Hao-Hai)/Cpw-(Wao-Wai)*(Tco-Two))/Ran;
  if(LG==0.) /* for L/G=0 return asymptote */
    return(asym);
  if(LG>=asym) /* for L/G>asymptote return 99.99 */
    return(99.99);
  K1=0.; /* lower bound for KaV */
  K2=999.; /* upper bound for KaV */
  dY=1./n; /* differential step size */
  for(i=0;i<25;i++) /* bisection search method */
    {
    KaV=(K1+K2)/2;
    LG2=LG+Wai-Wao; /* exit L/G is unknown */
    Ha=Hai; /* begin integration at air inlet */
    Wa=Wai;
    for(j=0;j<n;j++) /* step upward through the fill */
      {
      for(k=0;k<4;k++)
        {
        H=Ha;
        W=Wa;
        if(k)
          {
          H+=dY*A[k-1]*dH[k-1];
          W+=dY*A[k-1]*dW[k-1];
          }
        LG1=LG2+W-Wai; /* local L/G */
        Ta=fTdbhw(H,W); /* local Tdb */
```

```c
          Tw=Two+(LG2*Cpw*(Tco-Two)+H-Hai)/Cpw/LG1; /*
    local Tw   */
          Ws=fWsat(Tw); /* local Ws   */
          if(Tw>=Tho) /* check for overshoot */
            goto overshoot;
          dW[k]=LG*KaV*(Ws-W)/(1.+W);
          dH[k]=LG*KaV*((Ws-
    W)/(1.+W)*(Hwo+Cpg*Ta)+Lewis*Cpa*(Tw-Ta));
          }
       for(k=0;k<4;k++) /* last step of Runge-Kutta */
          {
          Ha+=dY*B[k]*dH[k];
          Wa+=dY*B[k]*dW[k];
          }
       }
    Wao=Wa; /* exit humidity is now known */
    if(Tw<Tho) /* bisection comparison */
       K1=KaV;
    else
       K2=KaV;
    continue;
    overshoot: /* overshoot occurrs if KaV is too large
    */
    K2=KaV;
    }
  return(KaV);
  }

int main(int argc,char**argv,char**envp)
  {
  double App,KaV,LG,Ran,Rh,Tco,Tdb,Tho,Twb;
  printf("\n");
  printf("                   computed KaV/L <-more nearly
    exact>\n");
  printf("Twb Ran App  L/G Merkel Marcel Rh= 50 Rh= 75
    Rh=100\n");
  for(Twb=50;Twb<=60;Twb+=10)
     {
     for(Ran=20;Ran<=30;Ran+=10)
        {
        for(App=10;App<=20;App+=10)
           {
           Tco=Twb+App;
           Tho=Tco+Ran;
           for(LG=.5;LG<=2;LG+=.5)
              {
              printf("%3.0lf %3.0lf %3.0lf
    %4.2lf",Twb,Ran,App,LG);
              KaV=KMerkel(Twb,Ran,App,LG);
              if(KaV<99)
```

```
          printf(" %6.3lf",KaV);
        else
          printf(" ******");
        KaV=KMarcel(Twb,Ran,App,LG);
        if(KaV<99)
          printf(" %6.3lf",KaV);
        else
          printf(" ******");
        for(Rh=50;Rh<=100;Rh+=25)
          {
          Tdb=fTdbwrh(Twb,Rh);
          KaV=KExact(Tdb,Twb,Ran,App,LG);
          if(KaV<99)
            printf(" %6.3lf",KaV);
          else
            printf(" ******");
          }
        printf("\n");
        }
      }
    }
  return(0);
  }
```

The output of the program is listed below.

```
                    computed KaV/L <-more nearly exact>
Twb Ran App L/G Merkel Marcel Rh= 50 Rh= 75 Rh=100
 50  20  10 0.50  2.325  2.219  2.330  2.287  2.256
 50  20  10 1.00  5.142  5.184  5.399  5.336  5.287
 50  20  10 1.50 ****** ****** ****** ****** ******
 50  20  10 2.00 ****** ****** ****** ****** ******
 50  20  20 0.50  1.102  1.062  1.106  1.094  1.086
 50  20  20 1.00  1.456  1.428  1.486  1.472  1.463
 50  20  20 1.50  2.462  2.552  2.658  2.634  2.616
 50  20  20 2.00 ****** ****** ****** ****** ******
 50  30  10 0.50  2.936  2.804  2.929  2.884  2.851
 50  30  10 1.00  7.586  7.687  7.974  7.805  7.805
 50  30  10 1.50 ****** ****** ****** ****** ******
 50  30  10 2.00 ****** ****** ****** ****** ******
 50  30  20 0.50  1.432  1.384  1.438  1.425  1.416
 50  30  20 1.00  2.011  1.986  2.063  2.047  2.035
 50  30  20 1.50  4.224  4.604  4.878  4.760  4.731
 50  30  20 2.00 ****** ****** ****** ****** ******
 60  20  10 0.50  1.707  1.656  1.753  1.720  1.696
 60  20  10 1.00  2.639  2.635  2.773  2.734  2.705
 60  20  10 1.50 ****** ****** ****** ****** ******
 60  20  10 2.00 ****** ****** ****** ****** ******
 60  20  20 0.50  0.835  0.817  0.858  0.849  0.843
 60  20  20 1.00  1.012  1.004  1.054  1.044  1.037
```

```
60  20  20  1.50  1.334   1.359   1.426   1.413   1.404
60  20  20  2.00  2.263   2.557   2.689   2.665   2.647
60  30  10  0.50  2.124   2.060   2.169   2.136   2.111
60  30  10  1.00  3.485   3.491   3.658   3.616   3.586
60  30  10  1.50  ******  ******  ******  ******  ******
60  30  10  2.00  ******  ******  ******  ******  ******
60  30  20  0.50  1.073   1.051   1.102   1.093   1.086
60  30  20  1.00  1.347   1.343   1.407   1.396   1.389
60  30  20  1.50  1.904   1.967   2.060   2.046   2.036
60  30  20  2.00  4.193   5.420   5.854   5.854   5.854
```

Appendix G: Generalized Runge-Kutta Function

We will use the Butcher tableau form to define each of the methods. The following C code implements any of these:

```
/* Butcher tableau and working vector indexing macros
    */
#define a(i,j)  BT[(steps+1)*i+1+j]
#define b(i)    BT[(steps+1)*steps+1+i]
#define c(i)    BT[(steps+1)*i]
#define k(i,j)  W[n*(i+1)+j]
void RungeKutta(double*BT,int steps,void dYdX(double,
    double*,double*),double*X,double dX,double*Y,int n)
  {
  int i,s,step;
  static int m;
  static double*W;
  if(W==NULL||n*(steps+1)>m)
    {
    if(n*(steps+1)>m&&m>0)
      free(W);
    m=n*(steps+1);
    W=calloc(m,sizeof(double));
    }
  dYdX(*X,Y,&k(0,0));
  for(step=1;step<steps;step++)
    {
    for(i=0;i<n;i++)
      {
      W[i]=Y[i];
      for(s=0;s<step;s++)
        W[i]+=dX*a(step,s)*k(s,i);
      }
    dYdX(*X+dX*c(step),W,&k(step,0));
    }
  for(i=0;i<n;i++)
    for(s=0;s<steps;s++)
      Y[i]+=dX*b(s)*k(s,i);
  *X=*X+dX;
  }
```

This concise function allocates and preserves a working array in which to store the intermediate results. This array is expanded when necessary. When we are evaluating the performance of the different methods, we do not want to allocate and de-allocate an array every time the function is called, as this can take more time than the calculations and bias the results. This code can be found in the on-line archive in a file named RKcomparison.c. This is already compiled and will run on any version of the Windows® operating system, but there is also a batch file to recompile it, named: _compile_RKcomparison.bat. All of the tableaus can be found in this source code.

Appendix H: Finite Difference Operators

There are countless references to which you might look for guidance regarding finite difference operators. If you weren't already familiar with these concepts, you would not have gotten so far along with this book. There is one text that stands above the rest when it comes to reminding you of what you should have remembered, filling in the gaps, and taking you farther into the material and that is the *Handbook of Mathematical Functions* by Abramowitz and Stegun.[52] You can download a PDF of this 470 page treasure trove here:

http://people.math.sfu.ca/~cbm/aands/abramowitz_and_stegun.pdf

You will find everything you need to be reminded about finite difference operators in Chapter 25, "Numerical Interpolation, Differentiation, and Integration." I will only include one of the many such formulas available in this landmark reference, as I wouldn't want to deprive you of the experience of looking up the rest.

25.3.32 Biharmonic Operator

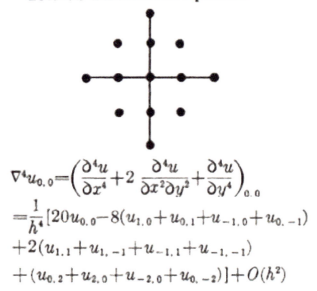

$$\nabla^4 u_{0,0} = \left(\frac{\partial^4 u}{\partial x^4} + 2\frac{\partial^4 u}{\partial x^2 \partial y^2} + \frac{\partial^4 u}{\partial y^4}\right)_{0,0}$$

$$= \frac{1}{h^4}[20u_{0,0} - 8(u_{1,0} + u_{0,1} + u_{-1,0} + u_{0,-1})$$
$$+ 2(u_{1,1} + u_{1,-1} + u_{-1,1} + u_{-1,-1})$$
$$+ (u_{0,2} + u_{2,0} + u_{-2,0} + u_{0,-2})] + O(h^2)$$

---

[52] Abramowitz, M. and I. A. Stegun, *Handbook of Mathematical Functions* first published by the National Bureau of Standards as Technical Monograph No. 55. This very useful reference may be obtained free on-line as a PDF from several different web sites.

Appendix I: Processing and Checking Finite Element Models

There are several errors that can creep into a finite element model, including: unused nodes, coincident boundaries, overlapping boundaries, coincident nodes, elements having duplicate nodes, and degenerate elements. The program CHECK2D.c in folder examples\finite element\checker checks for all of these errors. Typical inputs are:

```
51 nodes
421.333 733.333
460.667 854.667
500 976
etc...
70 elements
2 3 4
8 9 10
26 27 28
etc...
```

Typical outputs are:

```
CHECK2D: check 2D elements
input file=star.2dv
reading nodes
  expecting 51 nodes
  allocating memory for nodes
  51 nodes found
  0≤X≤1000
  24≤Y≤976
reading elements
  expecting 70 elements
  allocating memory for elements
  70 elements found
checking for errors
  unused nodes... none
  coincident boundaries... none
  overlapping boundaries... none
  coincident nodes... none
  elements having duplicate nodes... none
  degenerate elements... none
elements are clockwise
```

This code contains several useful functions, including: polygon area and inside polygon test.

Appendix J: Solving Large Systems of Linear Equations

There are several methods available for solving large systems of linear equations. These are often sparse and have certain characteristics, including: being positive definite and occasionally symmetric. The two we will consider here are Successive Over Relaxation (SOR) and Conjugate Gradient (C-G). The SOR method is trivial to implement: simply initialize to something reasonable and then iterate until the changes become small. This method could be called: *Plug-and-Chug*. A relaxation factor of 1.5 is typical. The following code implements this calculation:

```
void SuccessiveOverRelaxation()
  {
  int i,iTx,j;
  double dT,dTx,Told,Tnew;
  dTx=0;
  iTx=0;
  converged=1;
  for(i=0;i<Nn;i++)
    {
    Told=Tn[i];
    Tnew=Bn[i];
    for(j=0;j<Na[i];j++)
      Tnew-=Tn[Ia[Ma*i+j]]*An[Ma*i+j];
    Tnew/=Dn[i];
    Tnew=(1.-Relax)*Told+Relax*Tnew;
    Tn[i]=Tnew;
    dT=fabs(Tnew-Told);
    if(dT>dTmax)
      converged=0;
    if(dT<=dTx)
      continue;
    iTx=i;
    dTx=dT;
    }
  }
```

The C-G method was developed by Hestenes and Stiefel.[53] It is also an iterative method. The approach is best presented in terms of vectors. The basic problem A[]X[]=B[] can be rearranged to produce a residual: B[]-A[]X[]=R[], where R[] is the residual obtained with each estimate of the solution, X[]. The sum of the squares of the residual is $R^2 = R^T[]R[]$, where $R^T[]$ is the transpose of R[]. We define a step size along the direction between the previous and current estimates of the solution such that the R^2 is minimized. We then take another step that is orthogonal (perpendicular) the previous step and repeat the process.

---

[53] Hestenes, M. R. and Stiefel, E., "Methods Conjugate Gradients for Solving Linear Systems," *Journal of Research of the National Bureau of Standards*, Vol. 49, No. 6, Paper 2379, 1952.

If each step is orthogonal to all of the previous ones, there can be at most *n* steps, where *n* is the number of equations, because this is the total dimension of the space defined by the solution vectors. If the solution doesn't diverge due to round-off or ill condition of the matrix A[], the process must converge to the solution. This would be extremely burdensome and impractical if it weren't for the fact that an adequate solution is often achieved with only a few steps. The C-G method is truly remarkable in it's efficiency. Here's a simple function to implement the C-G method:

```
void ConjugateGradient(double*A,double*B,double*X,
   int*J,int n,int m)
  {
  int i,iter,k,l;
  double a,b,*D,dTq,q,*Q,r,*R,rTr;
  printf("begin Conjugate-Gradient\n");
  D=calloc(n,sizeof(double));
  Q=calloc(n,sizeof(double));
  R=calloc(n,sizeof(double));
  for(i=0;i<n;i++)
    X[i]=0.;
  for(rTr=l=i=0;i<n;i++)
    {
    r=B[i];
    for(k=0;k<m;k++,l++)
      if(J[l]>=0)
        r-=A[l]*X[J[l]];
    D[i]=R[i]=r;
    rTr+=r*r;
    }
  for(iter=0;;iter++)
    {
    for(dTq=l=i=0;i<n;i++)
      {
      for(q=k=0;k<m;k++,l++)
        if(J[l]>=0)
          q+=A[l]*D[J[l]];
      Q[i]=q;
      dTq+=D[i]*q;
      }
    a=rTr/dTq;
    for(i=0;i<n;i++)
      X[i]+=a*D[i];
    b=rTr;
    for(rTr=i=0;i<n;i++)
      {
      R[i]-=a*Q[i];
      rTr+=R[i]*R[i];
      }
    b=rTr/b;
    for(i=0;i<n;i++)
```

```
      D[i]=R[i]-b*D[i];
    if(iter%100==0)
      printf("iter=%li, rTr=%lG, a=%lG,
   b=%lG\n",iter,rTr,a,b);
    if(rTr<FLT_EPSILON)
      break;
  }
  free(D);
  free(Q);
  free(R);
}
```

This code along with sample inputs can be found in the on-line archive in the folder examples\matrix in file conjugate_gradient.c. This code also contains a function for the Gauss-Siedel method, which is just the SOR method with the relaxation parameter set to 1. Typical output for both methods is:

```
begin Gauss-Siedel
iter=0, rTr=2.16423
iter=1, rTr=0.666334
iter=2, rTr=0.253819
iter=3, rTr=0.10727
iter=4, rTr=0.0472994
iter=5, rTr=0.0211947
...
iter=18, rTr=5.77393E-007
iter=19, rTr=2.55549E-007
iter=20, rTr=1.13083E-007
begin Conjugate-Gradient
iter=0, rTr=4.50617, a=0.555556, b=0.450617
iter=100, rTr=0.00476629, a=0.0421095, b=1.22691
iter=200, rTr=0.000302086, a=0.00838571, b=1.24215
iter=300, rTr=1.40871E-005, a=0.0125829, b=1.19524
iter=400, rTr=9.45145E-006, a=0.00536786, b=1.18726
iter=500, rTr=2.41304E-006, a=0.010531, b=1.14972
iter=600, rTr=3.11957E-007, a=0.00717648, b=1.15689
iter=700, rTr=1.58797E-007, a=0.00515067, b=1.15158
```

I chose marginally ill-conditioned A[] and initial estimate X[]={0} for the C-G method just to illustrate the point that this method will converge eventually, regardless of the initial estimate. Most examples converge much faster than this, including the following.

Vectorization

Some matrix solution methods can be vectorized and some can't. The Gauss-Siedel and SOR methods don't lend themselves to vectorization, or if they do, there is negligible improvement in speed. Gauss Elimination (GE) and the Conjugate-Gradient (C-G) methods do vectorize quite well. Some machines and operating systems come with vector processing instructions, while on others, this can be implemented with software, usually written in assembler. Some graphics cards provide vector processing and have special functions that you can call to access this power. When you're working with very large matrices or solving many problems sequentially, this can be of great advantage.

I have included a program with source code to implement vectorization for these two methods. You can find it in the folder examples\matrix. This program (vector.c) allocates arrays (you can change the size), fills them with a simple problem (Laplace's Equation with Dirichlet boundary conditions), and solves it using the various methods. Program output looks like:

```
Test Banded Matrix Solvers
by Dudley J. Benton
Solving Laplace's Equation
with Dirichlet boundary conditions
grid points: 8x8=64

solving using Gauss-Seidel
iter=1, resid=0.5
iter=2, resid=0.1875
iter=3, resid=0.11377
iter=4, resid=0.0830078
iter=5, resid=0.0622158
iter=30, resid=0.000301372
iter=31, resid=0.000244635
error reduction 0.00048927

solving using SOR
iter=1 resid=0.75
iter=2 resid=0.166992
iter=3 resid=0.0967551
iter=4 resid=0.0674985
iter=5 resid=0.0345428
iter=13 resid=0.000773403
iter=14 resid=0.000147819
error reduction 0.000197091

solving using vectorized Hestenes-Stiefel
(Conjugate-Gradient)
iter=1 resid=2
iter=2 resid=0.875
iter=3 resid=0.455696
iter=4 resid=0.471959
iter=5 resid=0.307282
iter=6 resid=0.257967
iter=7 resid=0.136661
iter=8 resid=0.0796246
iter=9 resid=0.0366493
iter=10 resid=0.0129086
iter=11 resid=0.00130093
error reduction 0.00020064

solution
```

```
node    G-S      SOR        HSV
1       1        1          1
11      0.908066            0.90853  0.908622
21      0.631672            0.632514         0.632472
22      0.499389            0.5      0.499978
23      0.300216            0.300517         0.300591
64      0        0          0

Testing Gauss Elimination
1 1  1    1    1    1    1
1 2  4    8    16   32   2
1 3  9    27   81   243  6
1 4  16   64   256  1024 24
1 5  25   125  625  3125 120
1 6  36   216  1296 7776 720
vectorized
  scalar   vector
1 -264     -264
2 611.8    611.8
3 -511.583 -511.583
4 198.625  198.625
5 -36.4167 -36.4167
6 2.575    2.575
```

The source code contains vector instructions implemented in C, including: add, subtract, multiply, divide, dot product, and swap.

```
void vadd(double*v1,int incr1,double*v2,int
   incr2,double*v3,int incr3,int n)
{ /* v3=v1+v2 */
int i1,i2,i3,j;
for(i1=i2=i3=j=0;j<n;j++)
  {
  v3[i3]=v1[i1]+v2[i2];
  i1+=incr1;
  i2+=incr2;
  i3+=incr3;
  }
}
void vsub(double*v1,int incr1,double*v2,int
   incr2,double*v3,int incr3,int n)
{ /* v3=v1-v2 */
int i1,i2,i3,j;
for(i1=i2=i3=j=0;j<n;j++)
  {
  v3[i3]=v1[i1]-v2[i2];
  i1+=incr1;
  i2+=incr2;
  i3+=incr3;
  }
}
```

```
void vmpy(double*v1,int incr1,double*v2,int
   incr2,double*v3,int incr3,int n)
  { /* v3=v1*v2 */
  int i1,i2,i3,j;
  for(i1=i2=i3=j=0;j<n;j++)
     {
     v3[i3]=v1[i1]*v2[i2];
     i1+=incr1;
     i2+=incr2;
     i3+=incr3;
     }
  }
double vdot(double*v1,int incr1,double*v2,int incr2,int
   n)
  { /* v1 dot v2 */
  int i1,i2,j;
  double d;
  for(d=i1=i2=j=0;j<n;j++)
     {
     d+=v1[i1]*v2[i2];
     i1+=incr1;
     i2+=incr2;
     }
  return(d);
  }
void vswp(double*v1,int incr1,double*v2,int incr2,int
   n)
  { /* vector swap */
  int i,i1,i2;
  double v;
  for(i1=i2=i=0;i<n;i++)
     {
     v=v1[i1];
     v1[i1]=v2[i2];
     v2[i2]=v;
     i1+=incr1;
     i2+=incr2;
     }
  }
```

The scalar C-G method has already been discussed and can be found in conjugate_gradient.c. The vectorized version for a span of 5 is listed below.

```
int Hest5(double*a1,double*a2,double*a3,double*a4,
   double*a5,double*b,double*x,double*r,double*p,
   double*ap,double*w,int nx,int
   nn,double*r1,double*r2)
  {
  int i,iter,nnx;
  double alpha,epsilon,pap,pr,rap;
  /* vectorized Hestenes-Stiefel (Conjugate Gradient)
```

```
   method for linear simultaneous equations */
if(nn<=nx||nx<3||nn<9)
   return(OTHER);
nnx=nn-nx;
alpha=0.;
vset(0.,p,1,nn);
vset(0.,r,1,nn);
i=1+vmib(a3,1,nn);
if(fabs(a3[i-1])<DBL_EPSILON)
   return(SINGULAR);
vdiv(b,1,a3,1,x,1,nn);
for(iter=0;iter<nn;iter++)
   {
   if(iter==0)
      {
      vset(0.,r,1,nn);
      vsub(r,1,b,1,r,1,nn);
      vmpy(&a1[nx],1,x,1,&w[nx],1,nnx);
      vadd(&r[nx],1,&w[nx],1,&r[nx],1,nnx);
      vmpy(a2,1,&x[nx],1,w,1,nnx);
      vadd(r,1,w,1,r,1,nnx);
      vmpy(a3,1,x,1,w,1,nn);
      vadd(r,1,w,1,r,1,nn);
      vmpy(a4,1,&x[1],1,w,1,nn-1);
      vadd(r,1,w,1,r,1,nn-1);
      vmpy(&a5[1],1,x,1,&w[1],1,nn-1);
      vadd(&r[1],1,&w[1],1,&r[1],1,nn-1);
      }
   else
      vpiv(epsilon,ap,1,r,1,r,1,nn);
   i=1+vmab(r,1,nn);
   *r2=fabs(r[i-1]);
   if(iter==0)
      *r1=*r2;
   if(*r2<0.0005*(*r1))
      return(NOERROR);
   if(iter!=0)
      {
      rap=vdot(r,1,ap,1,nn);
      alpha=rap/pap;
      }
   vsmy(alpha,p,1,p,1,nn);
   vsub(p,1,r,1,p,1,nn);
   vset(0.,ap,1,nn);
   vmpy(&a1[nx],1,p,1,&w[nx],1,nnx);
   vadd(&ap[nx],1,&w[nx],1,&ap[nx],1,nnx);
   vmpy(a2,1,&p[nx],1,w,1,nnx);
   vadd(ap,1,w,1,ap,1,nnx);
   vmpy(a3,1,p,1,w,1,nn);
   vadd(ap,1,w,1,ap,1,nn);
```

```
      vmpy(a4,1,&p[1],1,w,1,nn-1);
      vadd(ap,1,w,1,ap,1,nn-1);
      vmpy(&a5[1],1,p,1,&w[1],1,nn-1);
      vadd(&ap[1],1,&w[1],1,&ap[1],1,nn-1);
      pap=vdot(p,1,ap,1,nn);
      if(fabs(pap)<DBL_EPSILON)
         return(SINGULAR);
      pr=vdot(p,1,r,1,nn);
      epsilon=-pr/pap;
      printf("iter=%i resid=%lG\n",iter+1,*r2);
      vpiv(epsilon,p,1,x,1,x,1,nn);
      }
   return(CONVERGENCE);
   }
```

The vectorized GE method is listed below. Both vectorized and scalar methods can be found in vector.c. The vectorizd version is implemented with conditional compilation directives (#ifdef... #else... #endif) to better illustrate the differences.

```
   int GaussVector(double*A,double*B,int n)
     { /* vectorized Gauss elimination with row and column
       pivoting */
     int i,j,k,ip,jp,*pivot;
     double a,b,p;
     if(n<1)
       return(OTHER);
     if(n==1)
       {
       a=A[0];
       if(fabs(a)<FLT_MIN)
         return(SINGULAR);
       B[0]/=a;
       return(NOERROR);
       }
     if((pivot=calloc(n,sizeof(int)))==NULL)
       return(OTHER);
     for(i=0;i<n;i++)
       pivot[i]=i;
     for(k=0;k<n-1;k++)
       {
#ifndef vectorized
     ip=k;
     jp=k;
     p=fabs(A[n*k+k]);
     for(i=k;i<n;i++)
       {
       for(j=k;j<n;j++)
         {
         a=fabs(A[n*i+j]);
         if(a>p)
```

```
            {
            ip=i;
            jp=j;
            p=a;
            }
          }
        }
#else
     jp=n*k+k+vmab(A+n*k+k,1,n*(n-k)-k);
     ip=jp/n;
     jp-=n*ip;
     p=fabs(A[n*ip+jp]);
#endif
     if(p<FLT_MIN)
        {
        free(pivot);
        return(SINGULAR);
        }
     if(ip!=k)  /* row pivot */
        {
#ifndef vectorized
       b=B[k];
       B[k]=B[ip];
       B[ip]=b;
#else
       vswp(B+k,1,B+ip,1,1);
#endif
#ifndef vectorized
       for(j=k;j<n;j++)
          {
          a=A[n*k+j];
          A[n*k+j]=A[n*ip+j];
          A[n*ip+j]=a;
          }
#else
       vswp(A+n*k+k,1,A+n*ip+k,1,n-k);
#endif
        }
     if(jp!=k)  /* column pivot */
        {
        j=pivot[jp];
        pivot[jp]=pivot[k];
        pivot[k]=j;
#ifndef vectorized
       for(i=0;i<n;i++)
          {
          a=A[n*i+k];
          A[n*i+k]=A[n*i+jp];
          A[n*i+jp]=a;
          }
```

```
#else
      vswp(A+k,n,A+jp,n,n);
#endif
      }
#ifndef vectorized
    for(i=k+1;i<n;i++) /* row elimination */
      {
      a=-A[n*i+k]/A[n*k+k];
      B[i]+=a*B[k];
      for(j=k+1;j<n;j++)
        A[n*i+j]+=a*A[n*k+j];
      }
#else
    for(i=k+1;i<n;i++) /* row elimination */
      {
      a=-A[n*i+k]/A[n*k+k];
      B[i]+=a*B[k];
      vpiv(a,A+n*k+k+1,1,A+n*i+k+1,1,A+n*i+k+1,1,n-k-
   1);
      A[n*i+k]=0.; /* must put 0s in lower triangle for
   vmab() to work */
      }
#endif
    }
  a=A[n*n-1];
  p=fabs(a);
  if(p<FLT_MIN)
    {
    free(pivot);
    return(SINGULAR);
    }
  B[n-1]/=a;
  for(k=1;k<n;k++) /* backsolve */
    {
    i=n-1-k;
#ifndef vectorized
    b=0.;
    for(j=i+1;j<n;j++)
      b+=A[n*i+j]*B[j];
#else
    b=vdot(A+n*i+i+1,1,B+i+1,1,n-i-1);
#endif
    B[i]=(B[i]-b)/A[n*i+i];
    }
  for(i=0;i<n;i++) /* column unpivot */
    A[i]=B[i];
  for(i=0;i<n;i++)
    B[pivot[i]]=A[i];
  free(pivot);
  return(NOERROR);
```

}

Besides replacing the for() loops with vector function calls, the only other change is that you must zero-out the lower part of the diagonal matrix in order for the vectorized pivot search to work properly. There's a comment on that line in the code. The vectorized pivot searches the entire bottom of the matrix in a single sweep. The scalar pivot only searches the lower right part of the matrix. If you don't put zeroes in lower part after you eliminate each row, the vector search will include these values in the search. It's not necessary to put zeroes in the lower part when performing the scalar search.

Appendix K: Triangular Element Generating Program

I have also included a small program (elem3.c) that will generate triangular elements from a closed polygon. It's very simple to use and all the files—including the source code—can be found in the on-line archive in the folder examples\grid generator. A typical polygon file looks like this:

```
POLY P1
0.510942647 -0.0510424392
0.549024467 -0.0889642514
0.600066906 -0.100004779
0.650949338 -0.0889642514
0.800076464 0
0.700071685 0
etc...
END
```

Which plots up with TP2 like this:

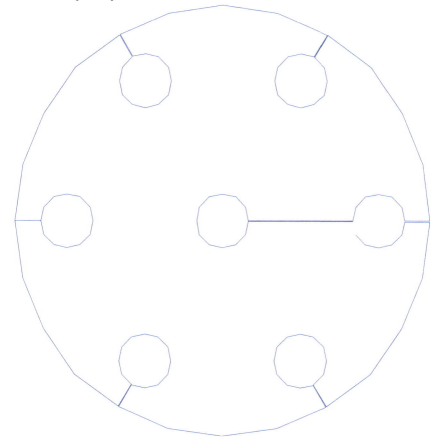

You have to split the lines that go into the bolt holes and then come back out to the edge. After processing it looks like this:

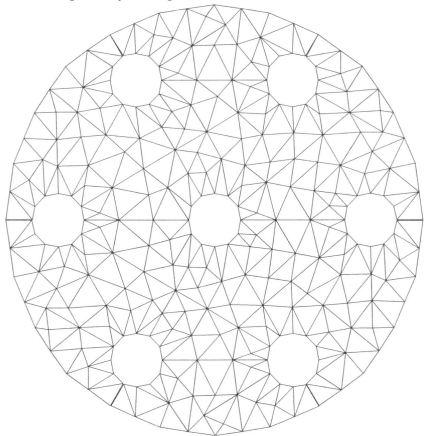

Typical output from the program is:
```
>elem3 lid.p2d
ELEM3/V2.00: Triangularization of a Closed Polygon
input file: lid.p2d
  scanning input file
  124 lines read
  122 boundary points found
  allocating memory
  reading input file
analyzing boundary polygon
  initializing points to boundary
  shortening sides
120 nodes, 124 elements
output file: lid.2dv
```

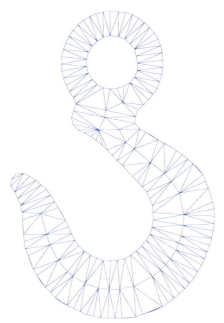

You can also control the size of the elements by specifying a target length as a second parameter after the input file name.

Appendix L: SIAM Paper
Applications of a Hybrid Derivative-Free Algorithm for Locating Extrema

I presented this paper at the Society of Industrial and Applied Mathematics, Southeastern Regional Seminar, held in Cullowhee, North Carolina, April 12-13, 1991. It's astounding how far computers have come in the past 27 years. Microprocessors are 100 times as fast and RAM has grown by a factor of 16. Disk space has increased ten-fold ($2^{10}=1024$). The test cases described herein are included in the online archive. They are still of interest, although there is less motivation to develop efficient methods with faster processing.

Abstract

Applications of a hybrid derivative-free algorithm for locating extrema of nonlinear functions of several variables based on Broyden's method is presented in which the problems of starting values and extraneous entrapment are addressed. The principal intended application of the algorithm is to find solutions to simultaneous nonlinear equations. The main objective of the algorithm is to minimize the number of function evaluations for problems where the equations are computationally intensive or partial derivatives cannot be determined analytically. Four examples drawn from diverse fields are given for illustration. Comparisons are made to the Newton-Raphson, conjugate-gradient, and steepest-descent methods.

Nomenclature

A=rectangular matrix having M columns and N rows
B=column matrix having M elements
F=a function of several variables
M=the number of residuals ($M \geq N$)
N=the number of unknowns
R=residual column matrix having M elements
X=unknown column matrix having N elements
superscript
T=matrix transpose
subscripts
N=new or current value
O=old or previous value

Introduction

Many practical problems can be cast into the form of a search for extrema of a function of several variables. A common function is the sum of squared residuals, in which case the extrema of interest are the roots of simultaneous equations. Methods abound which require knowledge of the partial derivatives. Many of these derivative-based methods can be adapted by using finite differences to solve problems where the partial derivatives cannot be analytically determined. Such implementations are impractical when the function is computationally intensive.

Derivative-based methods such as the Newton-Raphson discard at each step all information previously learned about the behavior of the function except the current location. Even the Conjugate-Gradient method when applied to nonlinear problems may only preserve one previous direction of search. When the function evaluation is computationally intensive it is essential that as much information as possible about the behavior of the function learned from previous evaluations be preserved and utilized.

Broyden's method is very attractive when considered from this perspective. It does not require knowledge of the partial derivatives, nor does it attempt to compute them directly. Furthermore, Broyden's method preserves all of the information learned about the behavior of the function for the last $N+1$ steps where N is the number of unknowns.

Four enhancements to Broyden's method were made to arrive at the present algorithm: a method for selecting starting values, step length control, hybrid search algorithm, and a method for escaping from extraneous entrapment.

The Basic Method

Given a set of N unknowns represented by the column matrix X and a corresponding set of M residuals represented by the column matrix R, the least-squares function would be $F=R^T R$. The extrema of F occur at the locations where $MF/MX=0$. If the residuals, R, were linear functions of the unknowns, X, then the function, F, would be quadratic and its contours would plot as ellipsoids. This linear case could be described by Equations L.1 and L.2:

$$R = A^T X + B \qquad (L.1)$$

$$F = R^T R = X^T A A^T X + 2 B^T A^T X + B^T B \qquad (L.2)$$

where A is a rectangular matrix having M columns and N rows and B is a column matrix having M elements.

Broyden (1969) reasoned that A and B should be selected such that exact agreement would be preserved for the previous $N+1$ steps. Assuming that no two of the previous $N+1$ Xs are the same, there should be a unique solution to the resulting $M(N+1)$ equations for the elements of A and B. Ignoring for the moment how this sequence of Xs might be obtained, the matrices A and B can be sequentially updated using the following algorithm:

$$A_N = \frac{A_O + [(R_N - R_O) - A_O(X_N - X_O)](X_N - X_O)^T}{(X_N - X_O)^T (X_N - X_O)} \qquad (L.3)$$

$$B_N = R_N - A_N^T X_N \qquad (L.4)$$

where the subscripts N and O refer to *new* and *old* respectively—or the current step and the previous one. Equations L.3 and L.4 can be verified by substitution into Equation L.1 with the *new* and *old* subscripts added. If A and B are initialized to zero and $N+1$ unique starting values of X are selected, then after $N+1$ function evaluations and updates, matrices A and B will be uniquely defined and the search for a solution could proceed.

The interesting property of Equation L.3, which led Broyden to this selection is that, the change in A is only in the direction of the last step in X. That is, the only information about the behavior of the function, which is added to A at each step, is its variation along the current search direction. All of the information about the function in the *N-1* directions orthogonal to the current search direction remains intact; thus, it is a *rank-one* update method.

Broyden used this algorithm for obtaining and updating matrices A and B, along with Newton's method to search for the extrema. Thus in its original form, Broyden's is a quasi-Newton method (Morè and Sorensen (1984) discuss Newton and quasi-Newton methods in some detail.). The following calculus can be applied to the matrices in order to illustrate this procedure.

$$\frac{\partial R}{\partial X} = A^T \qquad (L.5)$$

$$R_O = A_O^T (X_O - X_N) \qquad (L.6)$$

Matrix A contains the partial derivatives of the residuals, R, with respect to the unknowns, X. Thus matrix A is the *Jacobian* of R with respect to X.

As indicated by Equation L.6, the gradient of the function lies along the direction AR; therefore, most any gradient search method could be implemented and updated using Broyden's method for determining the Jacobian. Ortega and Rheinboldt (1970) discuss on an analytical level a number of methods which could be applied at this point. Actual selection of a practical method, which will produce satisfactory stable results for a wide range of problems, is quite another matter.

The Modified Method

Nonlinear simultaneous equations may have no solution, one solution, or many solutions. The most helpful physical analogy is that of a relief map of the Earth's surface where the unknowns are latitude and longitude and the function is the elevation with respect to mean sea level. No conceivable practical method could hope to locate Mt. Everest or the Marianas Trench regardless of the starting values. While it is reasonable to search for local extrema, it is fortuitous to locate the global extremum—assuming one does exist. Given this analogy it is understandable that no practical algorithm can be expected to locate even a local extremum in every case. Fletcher (1987) discusses these and other problems related to locating extrema in more detail.

Selection of Starting Values

This geographical analogy illustrates the necessity of limiting the region to be searched for extrema. In the present algorithm, a minimum and maximum value for each element in X must be supplied. This not only provides an extent to the range of X, but it also serves as an indication of the scale. Any change in X, which is on the order of the machine precision when compared to the range of X, is considered negligible. One logical choice for the $N+1$ starting values of X would be the center plus the N evenly distributed surrounding values inside the hypercube defined by the specified range of X.

If the function at the central point is greater than at the surrounding points, then the first iteration would direct the search outside of the range of X. If this occurs the range is bisected such that the new center point is mid way between the previous center and the surrounding point corresponding to the least value of the function. If this bisection is unsuccessful after sufficient attempts so as to diminish the subrange of any element of X to the previously determined negligible level, the search is abandoned.

Step Length Control

The unmodified method often results in unstable iterations. Not only is it necessary to confine X to the specified range, it is also necessary to damp the iteration or, as in this case, apply a step length control algorithm. Ortega and Rheinboldt discuss several step length algorithms. The parabola method defined by the current location, one *close* point, and the next step prescribed by the unmodified method has proven to be as successful as any tested. Using the unmodified point as an outer limit on the step length arises from the observation that the unmodified method has a strong tendency to overshoot.

Hybrid Search Algorithm

Because Broyden's is basically a quasi-Newton method, the search proceeds in much the same direction as with *Newton-Raphson* (*NR*). In cases where the *NR* method would fail to locate an extremum, most likely Broyden would also. Broyden's method can also be viewed as a means by which to obtain the Jacobian (matrix A). If the Newton iteration is not successful, the Jacobian can be used to implement other methods. The method of *steepest descent* (*SD*) is more robust, but converges less rapidly than *NR*. When the *NR* iteration fails to result in a reduction of the residual, the direction of steepest descent is searched.

In the present algorithm, the *Conjugate-Gradient* (*CG*) method with the restart procedure recommended by Powell (1977) is also used to supplement the *NR/SD* iteration. The only information added to the Jacobian by Broyden's update is along the search direction. Information about the character of nonlinear functions in directions orthogonal to the search direction can be essential to locating extrema. The *CG* method provides a systematic procedure for searching other directions. In the present algorithm, the *NR*, *SD*, and *CG* methods are used alternately as each ceases to provide continual reduction of the residual.

Escape from Extraneous Entrapment

If N directions have been searched without further improvement, then either a local extremum has been found or extraneous entrapment has occurred. Whether the current location is a local extremum or a nuisance of finite-precision arithmetic can be partially determined by examining the history of matrix A. For nonlinear problems the character of A can change substantially as the search proceeds.

The unmodified Broyden update to A replaces the information along the direction of the current step—thus discarding the previous information along this direction. If an *old* copy of A is retained along with the *new* copy and the search direction indicated by the old A is away from that indicated by the new A (viz. the dot product of the column matrices is less than or equal to zero), then the iteration may have skipped over an inflection point. In this case a search is conducted along the direction connecting these two provisional new values of X.

Extraneous entrapment can sometimes be corrected by arbitrarily perturbing the solution away from the current location to see if it will return to the same point. After this perturbation has been attempted without success in N directions the procedure is abandoned.

Extension to Least Squares

In the case where $M>N$, Equation L.7 must be pre-multiplied by A. The simultaneous nonlinear equations are then solved in the least-squares sense. For most problems the stability of the method also improves when this multiplication is performed even in the case of $M=N$. Therefore, in the present algorithm it is done regardless of the values of N and M.

Comparison to Other Methods

The present derivative-free *enhanced* Broyden (*EB*) method was compared to the Newton-Raphson (*NR*) and Conjugate-Gradient (*CG*) methods. The results are listed in Table 1. All three methods have step-size control and for the test cases were required to obtain essentially the same solution. All three methods were given the same starting values (initial guess) so that there was no advantage of one over the other in these respects.

Table 1 lists the number of variables (independent unknowns and dependent residuals), the number of function evaluations, and relative performance. The relative performance is the number of CPU-seconds required for the *NR* divided by the number required for the particular method (thus, *NR* will always have a relative performance of 1.0).

Test Case 1

The first test case is a nonlinear constrained curve-fitting problem. The best fitting single branch of a hyperbola was sought which would not only agree with the data (in this case experimental film boiling droplet area as a function of time), but would also have asymptotic characteristics conforming to the observed phenomena. The resulting curve fit must have one and only one root. The root must lie outside the range of the data and the derivative must be infinite at that point. The problem is nonlinear because of the constraints and the form (a rational polynomial). The partial derivatives of the residual cannot be determined analytically as these result in yet another set of simultaneous nonlinear equations. This test case was selected as being typical from among a set of 125.

Test Case 2

The second test case is similar to a nonlinear unconstrained curve-fitting problem. The values of hydraulic conductivity and storativity (groundwater analogs of electrical conductance and capacitance) were sought which would best characterize a measured field response. A field test was conducted by pumping water from a well and measuring the change in the water table in a nearby well. An analytical expression for the ideal response of an aquifer contains these two unknown parameters, which must be selected so as to best agree with the measured response. This problem is nonlinear, however the partial derivatives of the residual can be computed analytically (see note * in the table). This test case was selected as being typical from among a set of 33.

Test Case 3

The third test case is the determination of four *calibration factors* (mass transfer and pressure drop coefficients characterizing a particular type of plastic media), which are needed to run a large finite-integral code (numerical model of a cooling tower). Forty-nine sets of field data were collected for this plastic media. What were sought are the calibration factors which when input to the model would best reproduce the measured results. The finite-integral code itself was repeatedly run to provide the residuals. Needless to say, this was a very computationally intensive process—one in which minimizing the number of function evaluations was crucial. This test case, which is actually a type of inverse mass transfer problem, was selected as being typical from among a set of 6.

Test Case 4

The fourth test case is the determination of 4 phase lags and 4 corresponding weights, which would best characterize the transient response of a dammed reservoir. A linear model was sought for the cross-sectionally averaged transient flow at a specific location (adjacent to a large power plant) within a reservoir bounded by two dams, which are used for peaking (i.e., they discharge water only during times of peak electrical demand). This linear model was to become part of a larger linear systems optimization code used for long-range planning and resource management. An existing dynamic fluid flow model was used along with historical dam operations to produce a target data set. This test case was selected as being typical from among a set of 4.

Discussion

The focus of these four test cases is not on many variables, but on non-analytically differentiable residuals and lengthy function evaluations. In each case there is some physical phenomenon, which provides the basis for the residuals. Because these test cases are based on physical phenomena, the approximate bounds on the solution are also known. In each case lengthy graphical or cumbersome numerical techniques exist for finding extrema. The advantages to using the present algorithm in these cases are convenience and speed.

Table 1. Comparison of Methods for Locating Extrema

test case	variables		function evaluations			performance index		
	N	M	N-R	C-G	E-B	N-R	C-G	E-B
1	4	50	417	462	76	1	0.8	1.9
2	2	47	35*	51*	61	1	0.8	3.1
3	4	98	137	298	14	1	0.5	10.4
4	8	8760	103	344	18	1	0.4	2.4

*Note: Functions calls reporting analytical partial derivatives are more time consuming than those, which do not.

In the first test case (fitting a hyperbola with constraints and later taking its derivative) has been done for years using hand-drawn curves and a drafting protractor. The second test case (determining hydraulic conductance and capacitance) has also been done for years by graphical means and more recently by asymptotic extension to separate the coupled influence of the unknowns. The third test case (determining calibration factors for mass transfer and pressure drop) has typically been done by assuming half of the unknowns to be the same as a similar media and computing the others by *trial-and-error*.

For these test cases the average performance of the *EB* method is about 4 times the *NR* and *CG* methods. As mentioned previously, these are not isolated examples, but *working problems* from a variety of fields, which were the impetus for developing the method. The *EB* method utilizes the best features of the *NR*, *CG*, and *SD* methods along with avoiding direct calculation of the Jacobian. The relative advantage of the *EB* method was most dramatic for Test Case 3 where the difference in runtime was a matter of days (on a 33MHz-80386/7 machine).

A two-variable problem is best suited to illustrate the searching procedure graphically. Figures 1 through 3 show the contours of the function in Test Case 2 and the first few steps in the search path for the *NR*, *CG*, and *EB* methods respectively. The Z-axis or the contours is percent total residual (in 20% intervals). The dark (dense dot) region is close to the extremum and the light (sparse dot) region is far from the extremum. This graphical format was selected in order to give a *bulls-eye* appearance.

In this case the *CG* essentially follows the gradient inward to the center of the *bulls-eye* (see the dark path line in Figure 2). The *CG* path is almost perpendicular to the contours as it crosses each one. The *SD* path if it were shown would differ little from the *CG*. The *NR* and *EB* paths differ markedly from the *CG* (compare the dark path lines in Figures 1 and 3 to Figure 2). The *NR* and *EB* methods reach the vicinity of the extremum (i.e., penetrate the darkest inner contour) in significantly fewer steps than does the *CG* method.

The hybrid implementation of the present method can be seen by comparing the second step in the *NR* and *EB* paths. The line connecting the second and third points on the *NR* line (Figure 1) is almost parallel to the contour next to it (i.e., this step is almost perpendicular to the gradient). The line connecting the second and third points on the *EB* path (Figure 3) is almost perpendicular to the contour (i.e., almost in line with the gradient at the point where it crosses the inner contour). This illustrates how the *EB* method checks the search direction corresponding to all three methods (*NR*, *CG*, and *SD*) to see which is more advantageous at a particular location.

Conclusions

Broyden's derivative-free method for solving nonlinear simultaneous equations has been presented along with four enhancements. These enhancements include: a method for selecting starting values, step length control, hybrid search algorithm, and a method for escaping from extraneous entrapment. A significant performance improvement over the Newton-Raphson and Conjugate-Gradient methods is shown for four test cases taken from varied fields. Part of this performance improvement is a consequence of the derivative-free method. The hybrid search algorithm used in this enhanced Broyden method further improves the performance by utilizing the strengths of three other methods (the Newton-Raphson, Conjugate-Gradient, and Steepest-Descent).

References

Broyden, C., "A New Method of Solving Nonlinear Simultaneous Equations," *Computational Journal*, Vol. 12, pp. 94-99, 1969.

Fletcher, R., *Practical Methods of Optimization*, John Wiley and Sons, New York, NY, 1987.

Morè, J. J. and D. C. Sorensen, "Newton's Method," *Studies in Numerical Analysis*, G. H. Golub, ed., The Mathematical Association of America, pp. 29-82, 1984.

Ortega, J. M. and W. C. Rheinboldt, *Iterative Solution of Nonlinear Equations in Several Variables*, Academic Press, New York, 1970.

Powell, M. J. D., "Restart Procedures for the Conjugate Gradient Method," *Mathematical Programming*, Vol. 12, pp. 241-254, 1977.

also by D. James Benton

3D Articulation: Using OpenGL™, ISBN-9798596362480, Amazon, 2021 (book 3 in the 3D series).

3D Rendering in Windows®: How to display three-dimensional objects in Windows® with and without OpenGL™, ISBN-9781520339610, Amazon, 2016 (book 1 in the 3D series).

3D Models in Motion Using OpenGL™, ISBN-9798652987701, Amazon, 2020 (book 2 in the 3D series.

A Synergy of Short Stories: The whole may be greater than the sum of the parts, ISBN-9781520340319, Amazon, 2016.

Azeotropes: Behavior and Application, ISBN-9798609748997, Amazon, 2020.

bat-Elohim: Book 3 in the Little Star Trilogy, ISBN-9781686148682, Amazon, 2019.

Combined 3D Rendering Series: 3D Rendering in Windows®, 3D Models in Motion, and 3D Articulation, ISBN-9798484417032, Amazon, 2021.

Complex Variables: Practical Applications, ISBN-9781794250437, Amazon, 2019.

Compression & Encryption: Algorithms & Software, ISBN-9781081008826, Amazon, 2019.

Computational Fluid Dynamics: an Overview of Methods, ISBN-9781672393775, Amazon, 2019.

Computer Simulation of Power Systems: Programming Strategies and Practical Examples, ISBN-9781696218184, Amazon, 2019.

Contaminant Transport: A Numerical Approach, ISBN-9798461733216, Amazon, 2021.

Curve-Fitting: The Science and Art of Approximation, ISBN-9781520339542, Amazon, 2016.

Differential Equations: Numerical Methods for Solving, ISBN-9781983004162, Amazon, 2018.

Evaporative Cooling: The Science of Beating the Heat, ISBN-9781520913346, Amazon, 2017.

Heat Exchangers: Performance Prediction & Evaluation, ISBN-9781973589327, Amazon, 2017.

Heat Recovery Steam Generators: Thermal Design and Testing, ISBN-9781691029365, Amazon, 2019.

Jamie2: Innocence is easily lost and cannot be restored, ISBN-9781520339375, Amazon, 2016-18.

The Last Seraph: Sequel to Little Star, ISBN-9781726802253, Amazon, 2018.

Little Star: God doesn't do things the way we expect Him to. He's better than that! ISBN-9781520338903, Amazon, 2015-17.

Living Math: Seeing mathematics in every day life (and appreciating it more too), ISBN-9781520336992, Amazon, 2016.

Lost Cause: If only history could be changed…, ISBN-9781521173770, Amazon, 2017.

Mass Transfer: Diffusion & Convection, ISBN-9798702403106, Amazon, 2021.
Mill Town Destiny: The Hand of Providence brought them together to rescue the mill, the town, and each other, ISBN-9781520864679, Amazon, 2017.
Monte Carlo Simulation: The Art of Random Process Characterization, ISBN-9781980577874, Amazon, 2018.
Nonlinear Equations: Numerical Methods for Solving, ISBN-9781717767318, Amazon, 2018.
Numerical Calculus: Differentiation and Integration, ISBN-9781980680901, Amazon, 2018.
Orthogonal Functions: The Many Uses of, ISBN-9781719876162, Amazon, 2018.
Overwhelming Evidence: A Pilgrimage, ISBN-9798515642211, Amazon, 2021.
Particle Tracking: Computational Strategies and Diverse Examples, ISBN-9781692512651, Amazon, 2019.
Plumes: Delineation & Transport, ISBN-9781702292771, Amazon, 2019.
Power Plant Performance Curves: for Testing and Dispatch, ISBN-9798640192698, Amazon, 2020.
Props, Fans, & Pumps: Design & Performance, ISBN-9798645391195, Amazon, 2020.
Remediation: Contaminant Transport, Particle Tracking, & Plumes, ISBN-9798485651190, Amazon, 2021.
ROFL: Rolling on the Floor Laughing, ISBN-9781973300007, Amazon, 2017.
Seminole Rain: You don't choose destiny. It chooses you, ISBN-9798668502196, Amazon, 2020.
Steam 2020: to 150 GPa and 6000 K, ISBN-9798634643830, Amazon, 2020.
Thermochemical Reactions: Numerical Solutions, ISBN-9781073417872, Amazon, 2019.
Thermodynamic and Transport Properties of Fluids, ISBN-9781092120845, Amazon, 2019.
Thermodynamic Cycles: Effective Modeling Strategies for Software Development, ISBN-9781070934372, Amazon, 2019.
Thermodynamics - Theory & Practice: The science of energy and power, ISBN-9781520339795, Amazon, 2016.
Version-Independent Programming: Code Development Guidelines for the Windows® Operating System, ISBN-9781520339146, Amazon, 2016.
The Waterfront Murders: As you sow, so shall you reap, ISBN-9798611314500, Amazon, 2020.

Printed in Great Britain
by Amazon

27586092R00175